The Book of Japanese Tea

バイリンガル日本茶BOOK

ブレケル・オスカル
Per Oscar Brekell

淡交社 | TANKOSHA

はじめに

日本茶の魅力を一言で述べるとするなら、一口含めば馥郁たる香りに心身が包まれ、まるで「山中幽玄に心遊ぶ」ということではないでしょうか。日本には、東山文化のころから意識された「市中の山居」という言葉があります。茶道を大成したと言われる千利休などの茶人たちは、都会にありながら山里の閑居を思わせる茶室を作り、精神を開放させるための空間としてきました。

私はこの言葉こそ、日本茶の魅力そのものを表していると感じています。たとえ都会の喧騒の真ん中にいても、一服の茶の中に深山幽谷の静けさを感じ、立ちのぼる爽やかな香りから山の茶園に思いを馳せることが出来るからです。

もちろん、日本茶にはそれ以上に魅力がたくさんあります。苦味、渋味、そして甘味は他のお茶でも楽しめる要素ですが、日本茶は「うま味」が強調されていることが特徴の一つです。近年、和食の認知度が上がるにつれて「Umami」という日本語は、世界に通じる言葉となりつつあります。

うま味の成分は和食だけでなく、他国の料理にも含まれている大事な味の要素ですが、うま味が楽しめる嗜好飲料は世界的にも少なく、まさに日本茶ならではの特徴と言えるのでしょう。

加えて、日本茶は冷水から熱湯までと淹れ幅が広く、その茶が持つ全ての味をバランス良く出すことも出来ますが、自分でコントロールして飲みたい方向の

Preface

If I were to describe the beauty of Japanese tea with only a few words, I would say that by taking just one sip, the sweet fragrance immerses heart and mind in tranquility, and lets us enjoy the subtle beauty and elegant simplicity of the nature that surrounds the tea gardens deep in the mountains.

In Japan, the saying "*shichu no san-kyo*", or "mountain dwelling in the city" has been a well known concept since the age of the Higashiyama culture. The great tea masters like Sen No Rikyu that shaped the Japanese tea ceremony (the way of tea), constructed tea houses, that even if located in a noisy urban environment, would carry our thoughts away to a quiet and secluded life deep in the mountains, thereby creating a space where our minds can be liberated from the strains and struggles of everyday life.

I believe that this saying very well summarizes the true beauty of Japanese tea. Even if you happen to be in the middle of a bustling metropolis, the aroma that rises from your cup invigorates you and brings about peace of mind and serenity as if you were standing in the middle of a tea garden surrounded by misty forest clad mountains.

Needless to say, the beauty of Japanese tea stretches far beyond that. Bitterness, astringency and sweetness are elements that you can find in other teas as well, but the strong emphasis on *umami* is a true characteristic of Japanese tea. The Japanese word "*umami*" is now incorporated in many languages and is a natural element not only in Japanese cuisine but in most other food cultures as well. It is however very rare to find it in beverages and I believe that this truly makes Japanese tea stand out when compared to tea from other countries.

お茶を淹れることが可能です。例えば、淹れ方によって、同じ茶葉を使いながら、うま味を強調したり、香りを強く引き出したりすることが可能なのです。つまり、相手の好みや季節にあわせた味わいに出来、「おもてなし」の意識を具現化するのに最適な飲料であると思います。

淹れ方によって香りや味が変化するというと、難しく聞こえるかもしれませんが、基本さえ覚えれば誰もが出来るようになり、日本茶を淹れる楽しみは無限大に広がります。それには、使い勝手がよく、日本茶を美味しく淹れられる道具として日本の急須の助けを借りるのが一番です。長い日本茶文化の伝統の中で磨き上げられてきた急須はお茶を淹れる人を導き、その理からは学ぶことも多いものです。

私は、日本の食文化やおもてなしの文化を内包している日本茶こそ、日本文化の集大成を味わえるものだと考えています。ここで忘れてならないのは、「文化」とは、ただ伝統を守るものではなく、現代の私たちの中に生きているものであり、昔の知恵を基礎にして常に進化していくものだということです。現在、我々が飲んでいるお茶は、昔の人たちが飲んでいたお茶と味も姿も違います。製茶技術の進歩のおかげで、現代の日本では鮮度感と豊かなうま味を感じられる理想的な緑茶を機械によって製造することが可能になりました。かつての一部の高貴な人たちだけしか良質な茶を楽しめなかった時代を経て、今ではごく一般の人でも高品質の茶を手に入れることが出来ます。ワインのモノポールやシングルモ

Adding to that, Japanese tea can be prepared in many ways using either cold or hot water. The steeping method can be adjusted both to bring out a perfect balance of all its taste elements, and also to highlight a certain taste or flavor. You can, for example, choose to either accentuate *umami* or bring out as much of the aroma as possible. In other words, the tea can be adjusted to anyone's taste or to the season and therefore it perfectly embodies the concept of Japanese hospitality, or *omotenashi*, which is about putting all your efforts into the comfort and wellbeing of your guest. A tea with taste and flavor changing depending on the steeping technique may sound difficult to handle at first, but once you learn the basics the possibilities are endless and you will most likely find enjoyment in the steeping process, too. In your quest for a good cup of tea, you will find a good ally in the Japanese teapot that is both easy to use and makes the tea taste better. Japanese tea ware not only leads the way to a better cup of tea, but these utensils, shaped and improved by a long history and a rich cultural heritage, also add new depth and beauty to your tea experience.

Since both Japanese food culture and the culture of *omotenashi* are embedded in Japanese tea, I dare say that it allows us to enjoy the essence of Japanese culture. Here it is important to remember that culture is not just about preserving the traditions of the past, but something that lives inside of us, and based on the wisdom of the past and our current actions constantly continues to evolve. The tea that we are drinking today is very different in taste, flavor and appearance from the tea we drank in the past. Thanks to the progress of tea manufacturing, today it is possible to produce Japanese tea that comes close to the ideal green tea, having the characteristic fresh aroma while at the same time being strong in *umami*. Also, we have seen the end of the days when only

ルトなどが持つような個性的な味わいや物語を求めて、21世紀の初頭に登場してきたシングルオリジンの日本茶を好む人も現れて来ました。ただ、海外では水質の違いや緑茶の味に不慣れな人が多いせいか、嗜好品としての魅力よりも健康効果が強調される傾向にあります。確かにお茶を飲むと健康になると私も信じていますが、日本茶は長寿になるために飲むものというよりも、人生を豊かにしてくれる存在だと思います。世の中には辛いことや、日々の生活においてもストレスの原因となる煩わしいこともあります。そんな時には一杯の美味しい日本茶で心を落ち着かせ、前向きになることが出来ます。喜ばしい出来事や楽しい思いは、お茶を飲みながら多くの方と分かち合うことも出来ます。日本茶のそのような魅力や可能性に触れたひとりが、私です。この本をきっかけに、少しでも多くの方の人生が日本茶で幸せになればと心より願っています。

日本茶の楽しい世界へようこそ。

the upper layer of society could afford tea of decent quality, and today everyone can enjoy the tea they like. Adding to this, just like single vineyard wine or single malt whiskey, more and more people are discovering the unique and varied taste experiences of single estate Japanese tea, something that appeared as recent as the beginning of the 21th century.

In countries other than Japan, perhaps due to relative unfamiliarity with green tea, and different water quality that does not do justice to the original taste, there is still a tendency to emphasize the health benefits rather than the taste and flavor of the tea. Of course, I too believe that drinking green tea promotes health. However, rather than drinking Japanese tea to prolong one's life, I would like to think of it as something that makes our time on earth richer and more enjoyable. Life is full of hardships, and we always encounter both foreseen and unforeseen troubles that cause stress in our everyday lives. But by drinking Japanese tea we get a moment of peace, and we are able to gain the strength we need to remain positive in the midst of all this. Even during good times we can share our feelings and deepen our understanding of each other over a cup of green tea. I sincerely hope that reading this book will encourage more people to lead rich and fulfilling lives with the subtle beauty of Japanese tea as a part it.

Welcome to the wonderful world of Japanese tea.

目次

CONTENTS

そもそも「日本茶」って何ですか？

日本では様々な種類のお茶が作られています。紅茶や、烏龍茶の製法に近い釜炒り茶や漬物のように作る後発酵茶もあります。しかし、これらは生産量も少なく、「日本茶」というよりも地方の特産品として紹介されるもので、日本で生産されているお茶のほとんどは緑茶です。緑茶を製造する際には、生葉を加熱し、酸化酵素の働きを失活させる必要があります。大別すると、熱した鉄製の釜などで炒って製造された釜炒り製緑茶と、水蒸気の熱を利用して製造された蒸し製緑茶の二種類となり、日本の主な緑茶は後者に当たります。それゆえ、日本の蒸し製緑茶に親しむことで、「日本らしさ」を味わえることになります。

蒸し製緑茶は、出来るだけ生葉が持つ自然な香味が失われないよう、なるべく成分の変化を起こさないように製茶されます。その点でも、紅茶などの発酵茶や釜炒り製緑茶と大きく異なり、他の国のお茶にはない味わいが楽しめます。

また、もう一つの特色として和食文化と同様、強い香りの華やかさよりも「うま味」を追求してきた歴史があります。日本茶の多くは海外産のお茶と比べて、表立った派手さはありませんが、好きになればなるほど秘められた魅力に気付かされます。日本茶が求める「鮮度感」と「うま味」は、日本料理にも通じるもので、日本独自の製法と環境によって生み出されたものであり、ゆえに日本茶は極めて日本らしいものだと言えます。爽やかな香りで気持ちを豊かにし、淹れ方によって変化がつけられる香味は、毎日楽しんでも味わい尽くせないでしょう。

What is "Japanese tea" ?

Japan produces many different kinds of tea and among them we find black tea, oolong tea, pan-fired green tea, and even fermented tea to some extent. However, the production volumes of these are comparably small and therefore they are usually referred to as local specialties or rare teas rather than "Japanese tea". In fact, most tea produced in Japan is green tea. When processing green tea, the fresh leaves are heated to deactivate enzymes, thereby preventing oxidation. This is mainly done either by pan-firing or by steaming, but most Japanese teas belong to the second group. Since steaming is essentially a Japanese method, these teas can indeed be said to make up a unique Japanese taste experience.

Japanese teas are processed so as not to loose the natural taste and flavor of the fresh leaves, changing the chemical compounds as little as possible. In this sense, Japanese tea differs a lot from both oxidized tea like black tea and pan-fired green teas. This enables us to enjoy tastes and flavors that we cannot elsewhere. Another attribute peculiar to Japanese tea, also to be found in Japanese food culture in general, is the pursuit of and the emphasis put on *umami*, rather than aroma. Perhaps Japanese tea lacks grandiose characteristics but after having learned to appreciate it, one gradually becomes aware of the hidden beauty wrapped in the neatly rolled leaves.

Since both freshness and savoriness are sought after and praised as important characteristics of tea, as well as Japanese cuisine in general, one could argue that the country's food culture is indeed also expressed in its tea. The fresh aroma will induce a feeling of richness, and since the taste and flavor changes according to the chosen steeping method, new discoveries can be made indefinitely.

日本茶の味わい
The taste elements of Japanese tea

うま味

和食の普及とともに「うま味」という単語が全世界で広がりつつありますが、うま味はまさに日本茶の大きな特徴の一つです。一般的にはうま味の強い一番茶は高級茶とされ、逆にうま味の少ないお茶は下級品とされがちなほど、日本茶の品質を左右する要素です。うま味を呈する成分は主にグルタミン酸とテアニンですが、これらは低温の水でもよく溶出されます。

苦味

苦味は言葉として良い響きではないかもしれませんが、コーヒーやビターチョコレートと同様に大事な味の要素です。苦味の成分にはカテキンやカフェインなどがありますが、これらの成分はお湯の温度が高いとより多く溶出されるので、苦味が強くなり過ぎないようにするには温度を下げる工夫がお茶を淹れる際のポイントのひとつです。

渋味

苦味と間違える人もいますが、苦味と違って五基本味（甘味、酸味、塩味、苦味、うま味）の一つではなく、ポリフェノール類が口腔の粘膜と結合されることによってドライな感覚を与える感触に近いものです。紅茶のテアフラビン等を中心に渋味の成分がありますが、緑茶のカテキン類が与える渋味もまた独特な感覚です。

甘味

柔らかい芽で作られた一番茶は特にテアニンというアミノ酸が甘味を演出します。ぬるいお湯や冷水でもよく溶出しますので、甘味を強調したければ低い温度で淹れるのがポイントです。とくに一煎目はお菓子などを合わせず、日本茶の自然な甘味をそのまま味わうと良いでしょう。テアニン以外にも遊離糖類が甘味を呈しますが、微量ですのでカロリーの摂取を気にする心配はありません。

Umami

Just like Japanese food, the word *umami* has also crossed borders and spread throughout the world. Usually described as the sensation of richness or savoriness in food, it is a taste element that makes Japanese tea truly unique. *Ichibancha*, tea from the first harvest, is rich in *umami* and considered to be of high quality, whereas tea that has less of it in general is considered to be of lower quality. In this way, *umami* is a determining factor when grading Japanese tea. Glutamic acid and L-theanine, the main compounds behind the *umami* sensation, dissolve easily even when the tea is steeped at a low temperature.

Bitterness

Perhaps bitterness sounds unappetizing just by itself, but it is an important taste element in tea, just as it is in coffee and dark chocolate. Caffeine and catechins play the main role here, and these compounds are more easily dissolved in hot water. Therefore it is important to cool down the water when steeping Japanese tea to prevent excess bitterness from taking over the show completely.

Astringency

Some confuse astringency with bitterness, but unlike bitterness, which is considered to be one of the basic five taste elements (together with sweetness, sourness, saltiness, and *umami*), astringency is actually not a taste element, but rather a tactile sensation, leaving us with a dry sensation in the mouth as the polyphenols attach to proteins in our saliva. In black tea, thearubigin and other polyphenols cause a similar sensation, but the astringency in green tea is of a slightly different kind.

Sweetness

Especially Ichibancha, which is made from soft and tender young leaves, is rich in L-theanine, the main compound responsible for the sweetness in tea. L-theanine dissolves easily even in lukewarm or cold water, and therefore sweetness can be accentuated by lowering the water temperature. Sweetness also dissolves faster than astringency and bitterness, which makes the first steeping rich in natural sweetness, and enjoyable without pairing it with any confections. Apart from L-theanine, some free sugars also add to the sweetness in tea but there is no need to worry about any calorie intake as it is close to zero.

香り | Aroma

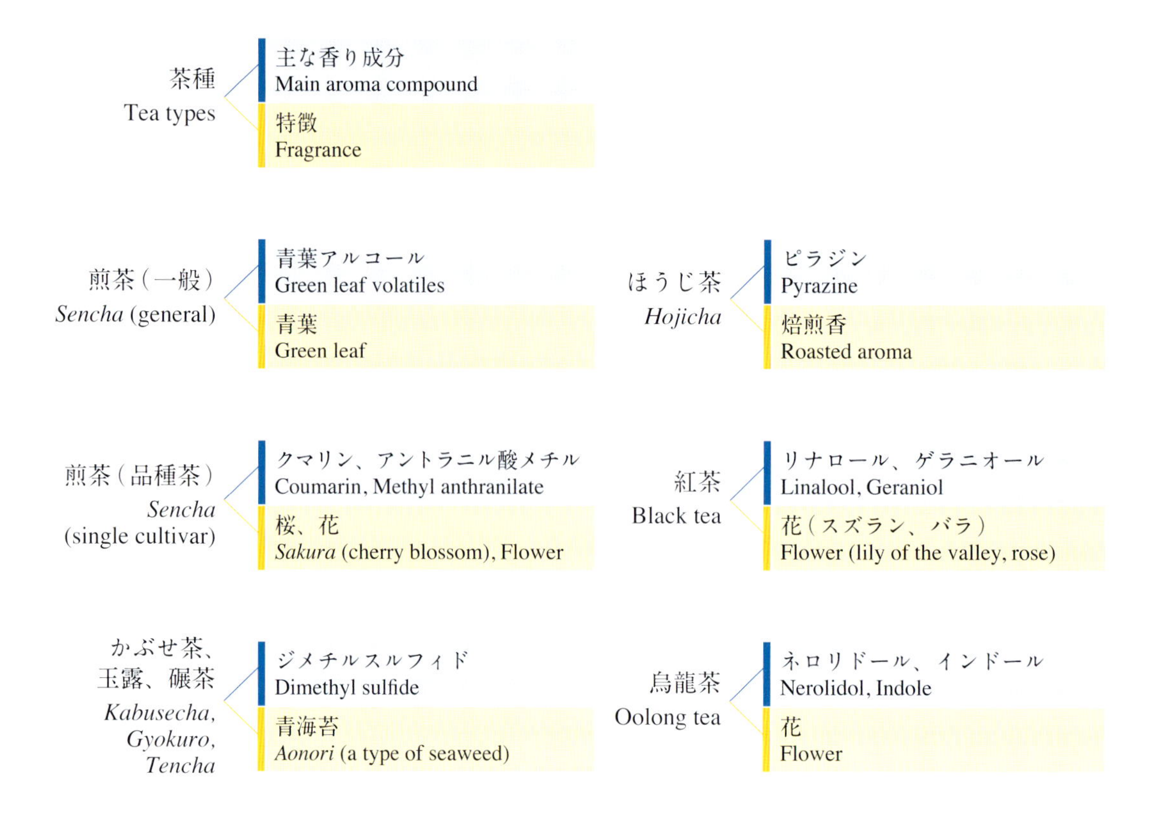

日本の蒸し製緑茶は、うま味を楽しむ嗜好品と認識されていますが、実は香りの面でも特色があります。紅茶や烏龍茶にはない、瑞々しく、青々とした森林のような香りを持ちます。特に新茶にはこの香りの元になる青葉アルコールが豊富に含まれています。

加えて品種によっては花や果実のような香りを発するものもあります。加熱や乾燥により香りの成分は変化し、揮発するため、日本茶らしい香り豊かなお茶を作るには、出来るだけ短時間で加熱や乾燥を効率的に行う必要があります。特に茶葉を焦がすことなく水蒸気を使って加熱し、酸化酵素の失活を行う「蒸す」工程は日本茶にとって重要であり、日本茶らしい香りの元を決める作業でもあります。

Japanese steamed green tea is often considered to be all about enjoying *umami*, but its refreshing forest-like aroma, a world apart from the flavor profile of black tea or oolong tea, also makes it stand out among other teas. Especially *Shincha* ("New tea", gently finish-fired *Ichibancha*) is rich in the green leaf volatiles that contribute to this characteristic aroma.

Adding to this, we also find the floral and fruity notes peculiar to certain cultivars of the tea plant. Aroma compounds are volatile and sensitive to heat, so in order to produce a tea that is rich in aroma, it is necessary to make the heating and drying processes as short and efficient as possible. To prevent oxidation, steam is used to de-enzyme (fix) Japanese tea so as not to scorch the leaves. Therefore steaming is a crucial step that gives the tea its typical Japanese aroma.

成分と効能 | Compounds and efficacy

日本の蒸し製緑茶は、形や香り、味も他国のお茶と異なっています。成分を調べると、その詳細がより明らかになります。渋味と苦味の元になるのはカフェインやカテキン類とされますが、カテキン類は抗酸化作用や細胞の癌化、老化につながる細胞の突然変異を抑制するのではと有望視されています。テアニンと呼ばれるアミノ酸の一種は日本茶に甘味とうま味を与え、摂取することで脳内にα波が発生して穏やかな気分になることが報告されています。前述のカフェインには覚醒作用があり、矛盾しているように思えますが、緊張やストレスを感じない覚醒効果が可能なのはこの両方の成分の働きのためでしょう。とはいえ、お茶は薬として飲むのではなく、日常的に美味しく味わうことによって気持ちに余裕が生まれ、日々の生活が豊かになります。成分の摂取にこだわりすぎず、習慣的に美味しく飲むことこそが日本茶との最も健康的な付き合い方でしょう。

Japanese steamed green tea is different from tea from other countries in both taste, flavor and appearance. The reasons for this become clear by looking at the chemical compounds in the tea. The main compounds behind bitterness and astringency are a group of polyphenols called catechins, and caffeine also contributes to the bitter taste of tea. Catechins are considered to be a powerful antioxidant and by protecting against gene mutation, also thought to slow down the aging process and prevent the formation of cancer cells. L-theanine, usually credited for the sweetness and *umami* in tea, is reported to promote alpha wave activity in the brain, making us feel calm and peaceful. As mentioned above tea also contains a stimulant, namely caffeine, which may seem contradictory but the sensation of being both alert and relaxed without feeling any stress or nervousness probably arises from the effects of both compounds taken together. That said, rather than consuming tea as a medicine, enjoying good tasting tea on an everyday basis provides us with breathing room and enrich our everyday life. Not being too particular about the intake of certain compounds, but enjoying good tea as a habit, is probably the healthiest approach to tea.

日本茶の種類 | Types of Japanese tea

茶の分類 | Tea classification

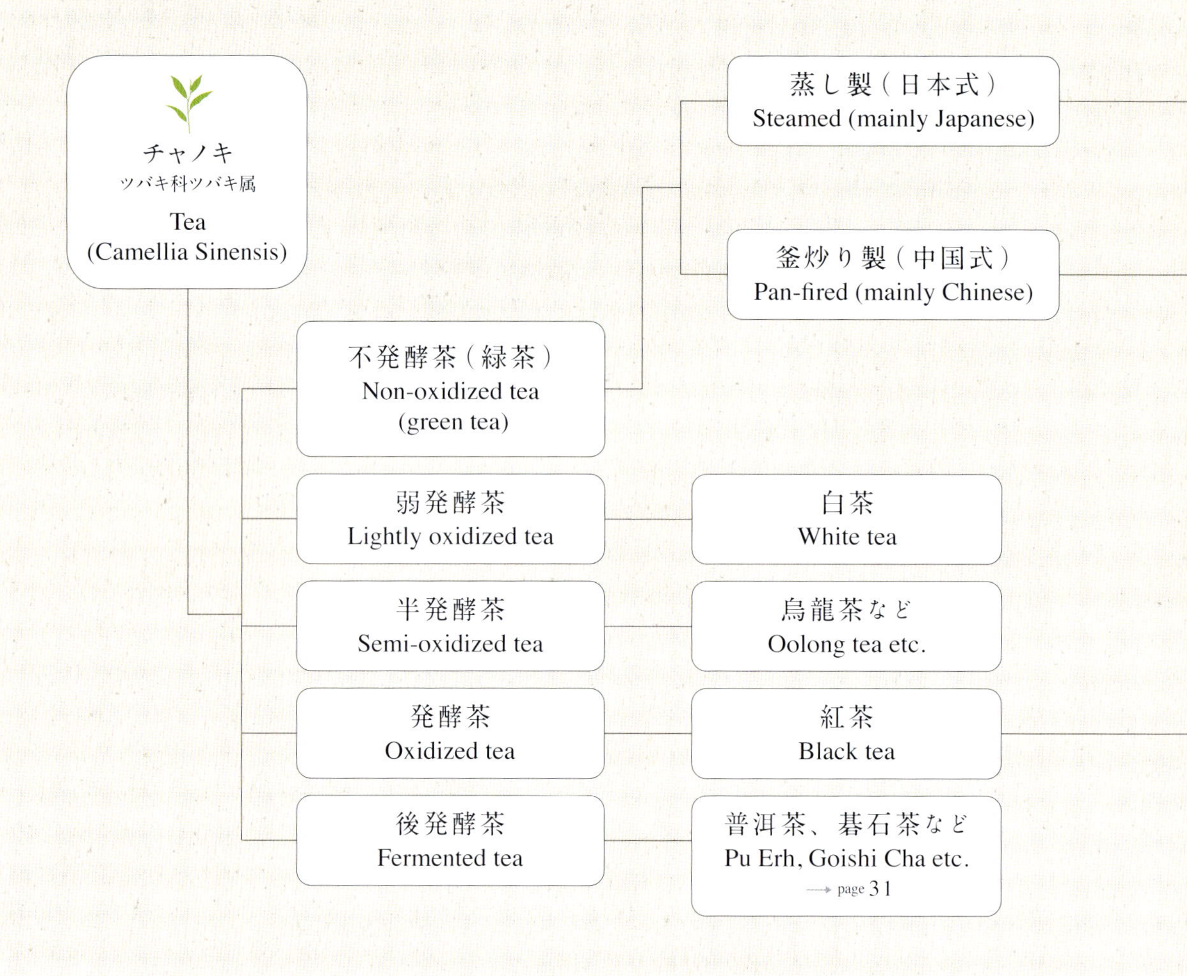

紅茶、烏龍茶、緑茶は全てチャノキ(カメリアシネンシス)の芽や茎を含んだ葉の部分で作られています。これらの香り、味、形の違いは製法によって生まれ、茶葉に含まれる酸化酵素をどの程度働かせたかによって、紅茶、烏龍茶、緑茶に分類されます。酸化酵素の働きを止めることを「殺青(さっせい)」と呼びます。殺青には様々な方法があり、同じ緑茶でもその方法によって結果が異なります。大別すると、加熱した釜などで炒って作る「炒り製」と水蒸気の熱を利用した「蒸し製」があります。前者は主に中国や台湾、後者は日本での製造法で、炒り製緑茶は香ばしい釜香と甘い香りであっさりした味わいのお茶、蒸し製緑茶は鮮度感が強く、ふくよかなうま味とさわやかな渋味のお茶になります。蒸し製緑茶をさらに細かく分類すると、覆いをかけずに栽培した「煎茶」と覆いをかけて日の光を遮って栽培した「被覆茶」に分かれます。煎茶の中で茶葉を長めに加熱して細かな形にしたものを「深蒸し煎茶」と呼び、玉露、かぶせ茶、碾茶(抹茶の原料)は被覆茶に該当します。上記の茶種以外には微生物の働きを活かして発酵させた普洱(プーアール)茶などもあり、チャノキの葉は製法によって様々な香味や姿になります。

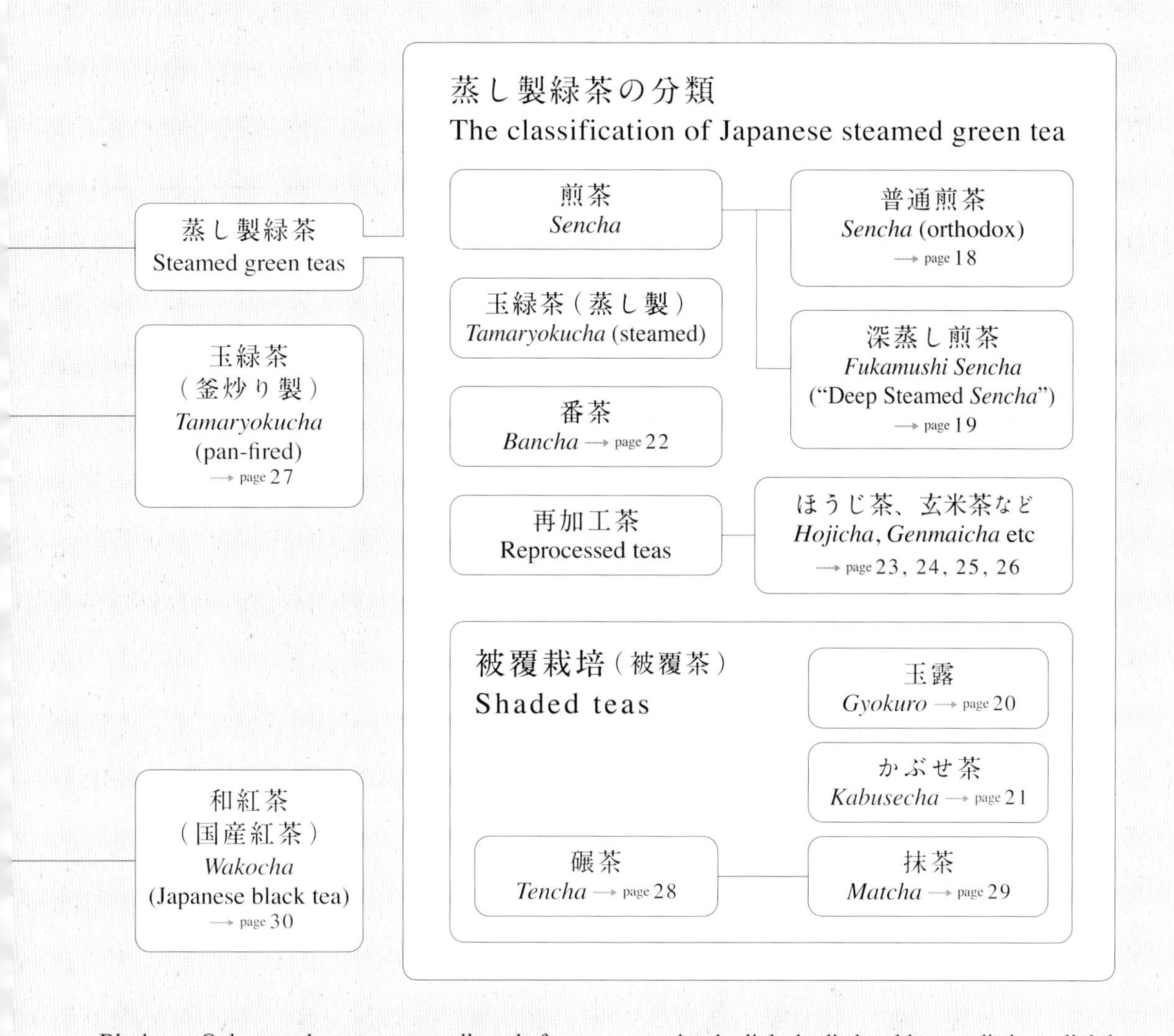

Black tea, Oolong and green tea are all made from the leaves and buds, and in some cases stems, of the tea plant (Camellia Sinensis). The difference in aroma, taste and appearance arise during processing, and depending on the degree of oxidation we end up with black, Oolong or green tea etc. Oxidation is prevented by deactivating enzymes in the tea leaf, and this process is called *sassei* in Japanese (written with the Chinese characters "kill" and "blue/green" and usually referred to as "fixing " or "kill-green" in English). The are several ways of deactivating the oxidation enzymes, and green tea will end up differently depending on the fixing method. Green tea can be divided into two main types: pan-fired green tea (usually heated in a pan or a tumbler) and steamed green tea. The former type is mainly produced in China and Taiwan, and the latter mainly in Japan. Pan-fired green teas tend to be light-bodied and have a distinct, slightly sweet, roasted aroma. Steamed green teas on the other hand have a characteristic fresh aroma, strong *umami* and a pleasant astringency. These teas can be further divided into non-shaded and shaded teas. In the case of shaded teas like *Gyokuro*, *Kabusecha* and *Tencha* (which is used to make *Matcha*), the plantation is covered before harvest to prevent exposure to direct sunlight. Among non-shaded teas we find *Sencha* and its cousin *Fukamushi Sencha*, which is steamed and processed for a longer time than *Sencha*, giving it a broken appearance. Apart from the above mentioned types there are also teas like Pu Erh (from China), fermented by the action of microorganisms. To sum it up, the very same Camellia Sinensis can be used in many ways to produce teas that are different in both taste, flavor and appearance.

煎茶 | Sencha (Orthodox)

美味しい淹れ方
How to steep Sencha ? → page 68, 72, 76

煎茶は日本茶を代表する茶種の一つですが、淹れ方による香味の変化が最も楽しめる茶種でもあります。使用する水が低温ならば甘味とうま味が強調された味わいで、逆に高温ならば苦味と渋味、そして香りが出やすくなります。正解はなく、基本を覚えたら試行錯誤しながら淹れてみると楽しいでしょう。味のバランスが取れたブレンドから個性豊かなシングルオリジンまで、淹れ方による香味の変化は幅が広く、キリがないほど奥が深い、面白い茶種です。

遮光をせずに露地で育てられ、産地や品種を活かした自然な香りが残るお茶になるので、ワインやシングルモルトのように飲み比べを楽しむにも最適です。

Sencha is one of the true classics from the repertoire of Japanese tea, and it stands out as a tea that can be steeped and enjoyed in a multitude of ways to evoke or accentuate different flavors. And in this sense, it is probably more flexible than any other tea. When using cold water, the natural sweetness of the tea is accentuated, and if water with a high temperature is used, bitterness, astringency and aroma are dissolved to a greater extent. There is no correct answer to the question of how you should steep your tea, so learning the logic behind tea steeping and then trying to apply it in different ways is by far the most enjoyable approach. Ranging from blends with a good balance in both taste and flavor, to unique single estate teas, which can both be enjoyed in a many different ways, the world of *Sencha* is both fascinating and endlessly deep. Since it is grown without shading, the natural flavor from the terroir and the cultivar remains, and this makes *Sencha* an excellent choice for the aficionado who likes comparing different teas in a way similar to wine or whisky tasting.

深蒸し煎茶 | Fukamushi Sencha ("Deep steamed Sencha")

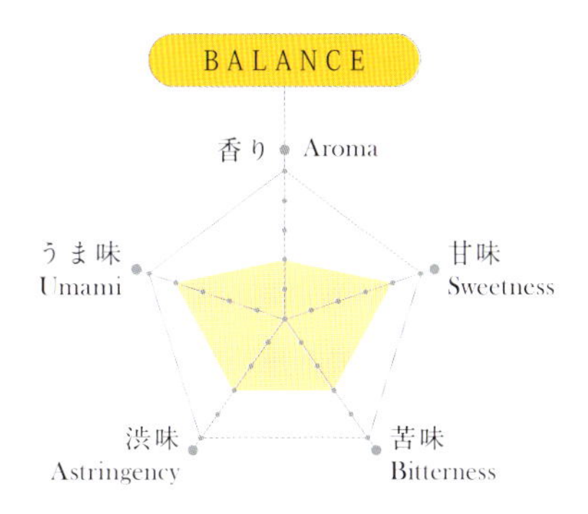

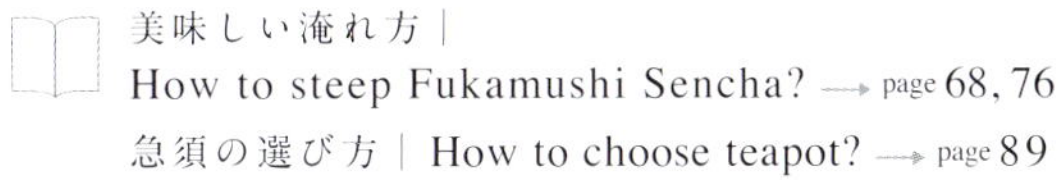

美味しい淹れ方 | How to steep Fukamushi Sencha? → page 68, 76
急須の選び方 | How to choose teapot? → page 89

深蒸し煎茶と呼ばれているお茶の共通点は、茶葉が細かく、水色は濁りがあって濃い緑色をしているところで、香りよりもコクのある味わいを楽しむお茶です。水蒸気を利用した加熱の時間を長めにして、揉む工程などで細かな形状にします。日照量が多い平坦地で栽培されたお茶は苦くなりがちですが、加熱時間を長くすると甘味を強調するペクチンという成分が溶出しやすくなり、マイルドに感じられます。この製法は1960年代に静岡県の中部で開発され、現在、日本で最も消費されている茶種です。

70℃くらいで淹れると雑味が少なく、美味しく味わえます。

Teas referred to as "*Fukamushi Sencha*"(usually translated as "Deep Steamed *Sencha*" or "Heavily steamed *Sencha*") all share the broken appearance and the cloudy liquor. Rather than aroma, it is sought after for its thick and rich taste. The steaming process is longer than for *Sencha* and it is during the rolling process that the tea gets its broken appearance. Tea grown in flat areas with long day light hours tend to become bitter, but by steaming the leaves for a longer time pectin, a compound that enhances sweetness, dissolves more easily and as a consequence we perceive *Fukamushi Sencha* as a mild tea. The processing method was developed in the 1960's in the central part of Shizuoka prefecture and nowadays it is the most widely consumed tea in Japan. To avoid any harsh taste or excess bitterness, it should be steeped in about 70°C (160°F) for the best result.

玉露 | Gyokuro

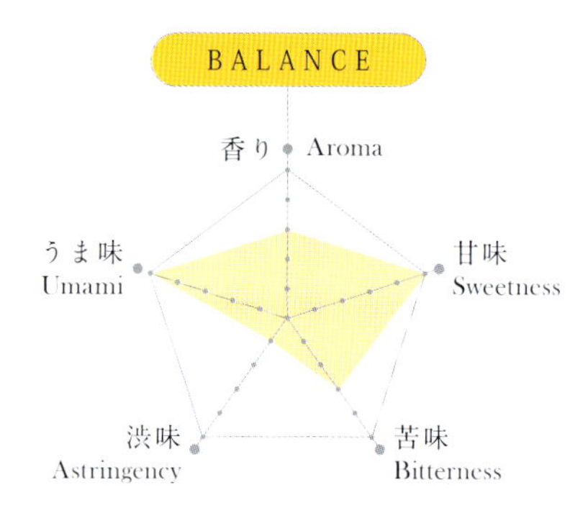

美味しい淹れ方
How to steep Gyokuro ? → page 72, 76, 80

日本茶はうま味を楽しむ嗜好品といわれますが、うま味を最も強く感じられるお茶のひとつが玉露です。生産量がわずか250トン程度と他の茶種と比べて少量しか生産されず、海外でも最高級茶として有名です。そのうま味には理由があります。玉露は煎茶とほぼ同じ製法でつくられますが、栽培方法が異なります。摘みとりの約3週間前に茶園に覆いをかけて光を遮ります。遮光により、うま味成分であるテアニンが渋味成分のカテキンなどに変化せず葉が成長していきます。加えて窒素分を多く含んだ肥料を十分に施すのが、玉露のうま味の秘密です。濃厚な味わいを楽しむお茶なので、小さな豆茶碗などで味わうのがベストです。水出しでも美味しく味わえます。

Enjoying *umami* is often considered to be an essential part of Japanese tea drinking, and *Gyokuro* is one of the teas where this is taken to the maximum. Only about 250 tons are produced every year, a small amount compared to other teas. The rich *umami* does not come about by chance. Although processed in a similar way to *Sencha*, it is grown under very different circumstances. About three weeks before the spring harvest, the tea plants are shaded using either a reed screen with straw spread on top, or a synthetic black cloth to prevent exposure to direct sunlight. Even as the leaves grow, shading prevents the *umami* compound L-theanine from turning into catechin, a group of polyphenols that make the tea taste bitter and astringent. This, and meticulous fertilizing of the tea garden, is the secret behind the strong *umami*. It is best steeped in low temperatures and, due to its almost overwhelming richness, best enjoyed in small quantities. *Gyokuro* also makes an excellent cold brew.

かぶせ茶 | Kabusecha

BALANCE

香り Aroma
甘味 Sweetness
苦味 Bitterness
渋味 Astringency
うま味 Umami

美味しい淹れ方
How to steep Kabusecha ? → page 68, 76

煎茶と玉露の両方の要素をある程度兼ね備えているため、両者の中間の存在と説明されます。名前の通り、茶園に覆いをかぶせて育てたお茶ですが、玉露よりも遮光期間が短く、4～10日が一般的です。玉露のように直立した樹形にするのではなく扇状の畝とし、「寒冷紗」と呼ばれる化学繊維の被覆材を、茶樹の上へ直に被せます。遮光することで渋味が減り、甘味が強く、青海苔のような独特な「かぶせ香」が生成され、煎茶とは異なった味わいになります。熱めのお湯で淹れても美味しく楽しめます。

Since *Kabusecha* possesses elements that can be found in both *Sencha* and *Gyokuro*, it is often described as something in between the two of them. “*Kabuse*” translates as “cover” in Japanese, and just as the term suggests, the tea plants are grown under shade, a cultivation method it shares with *Gyokuro*, although *Kabusecha* is shaded for a shorter time. Also, unlike *Gyokuro* where poles hold up a reed screen roof which is later covered by straw to let the tea plant grow freely, *Kabusecha* plantations are cut into hedges and a synthetic black cloth is put directly on the plants usually about 4-10 days before harvest. By shading the plantation like this, the tea becomes stronger in sweetness and weak in astringency. On top of this, we also find the characteristic *nori* sea weed like aroma which clearly makes it different from *Sencha*. *Kabusecha* tastes good even when steeped in relatively high temperatures.

番茶 | Bancha

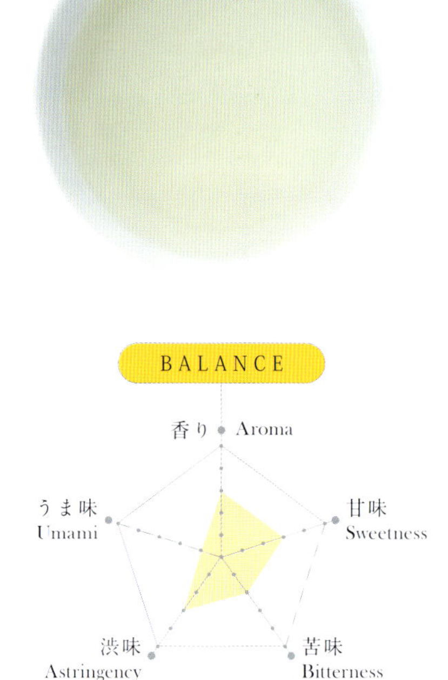

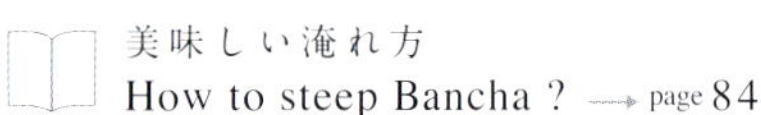
美味しい淹れ方
How to steep Bancha ? → page 84

番茶は茶種の中で最も紛らわしい名称の一つかもしれません。近代化の前には抹茶や煎茶などのように茶道に使われるお茶ではなく、庶民が日常的に飲んでいたお茶を指していました。それらの番茶は煮たり、炙(あぶ)ったりしてから乾燥させるなど様々な方法で作られましたが、現在の「番茶」は煎茶の製法で作られた二番茶以降のものを指すようになっています。昔ながらの番茶はほとんど姿を消し、地方番茶とも呼ばれます。番茶はうま味が強くなく、色や形も一番茶ほど整っていないため、下級なお茶とされがちですが、食後にもぴったりなさっぱりとした味わいのものが多いのでぜひ試してみてください。

Bancha is probably one of the most confusing terms in the Japanese tea world. In preindustrial days, it referred to teas consumed by the common people, as opposed to *Matcha* and higher quality leaf tea that was affordable only for those of higher ranks in society. That sort of *Bancha* could be processed in many different ways, e.g. boiled or roasted over an open fire before the drying process began. However, nowadays the term *Bancha* is mainly used to denote *Sencha* from the second harvest and beyond. The *Bancha* of old times has almost disappeared entirely, and is now referred to as "rural area *Bancha*". Since *Bancha* is not strong in *umami*, and does not match *Ichibancha* in color and appearance either, it is often thought of as a low grade tea, but a light-bodied *Bancha* can be the perfect choice after a meal as it both clears the palate and refreshes the mind.

京番茶（地方番茶） | Kyobancha

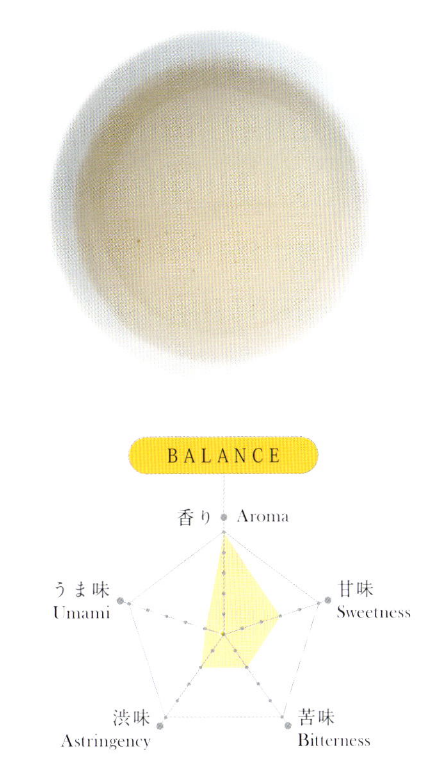

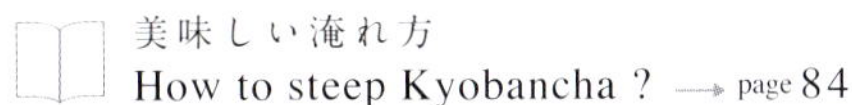
美味しい淹れ方
How to steep Kyobancha ? → page 84

京番茶は「炒り番茶」などとも呼ばれている京都の伝統的なお茶で、一般的な番茶（22頁参照）と違い、軽くてかさばる大きな茶色の葉をしています。初めて見ると、日本のお茶だと思わない人もいることでしょう。京都では伝統的な玉露や碾茶（抹茶の原料）の製造後に、茶の樹を膝下くらいの高さで刈り落とします。その刈り落とした枝葉で京番茶は作られます。蒸してから揉まずに乾燥させ、保存した後に炒ってから袋詰めをします。スモーキーな香りが強く、現代では珍しく感じられると思いますが、昭和の初期まで各地で親しまれていた番茶（庶民の茶）は、このお茶に近いものが多かったと考えられます。

This tea from Kyoto also goes by other names, such as “*Iribancha*” (roasted *Bancha*). Its light bulky brown appearance makes it worlds apart from ordinary *Bancha* (see p.22), even to the degree that unfamiliar Japanese would mistake it for a tea from another country. After leaves for *Gyokuro* and *Tencha* (the base for *Matcha*) have been picked, the tea plants are cut down to just below knee height to renew the plants for the next year’s first harvest, and it is all the leaves and branches that are cut off during this step that are used for *Kyobancha*. The leaves and branches are steamed, dried without rolling, stored and then finally roasted before being sent to the market. With its distinct smoky aroma and unusual appearance, many think of it as a rare tea, but until the 1920’s and 30’s many teas similar to *Kyobancha* were probably consumed as *Bancha* by the common people.

ほうじ茶 | Hojicha

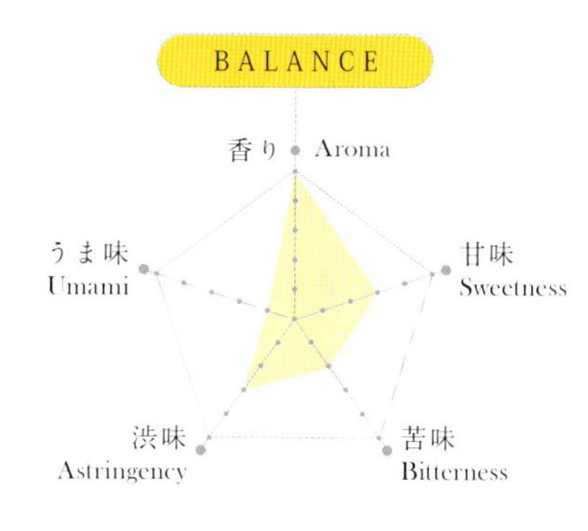

美味しい淹れ方
How to steep Hojicha ? → page 84

ほうじ茶は主に緑茶を焙煎することによって作られ、一般的には二番茶以降の茶が原料となります。焙煎の香ばしい香りがその魅力です。同じほうじ茶でも原料によって香味が変わります。特に良質な茎が多いほうじ茶は香り高くクリアな味わいとなり、原料中の葉の比率が増えると渋味もしっかりと感じられ、力強い味になります。日本では食後に好まれます。もし古くなった煎茶が自宅にあれば、捨てずにフライパンやほうじ器などで茶葉を軽く炒って、自分流のほうじ茶を作ってみましょう。ほうじ香が周辺に広がり、それも味わいの一つとなります。

Hojicha is made by roasting steamed green tea, and in general tea from the second harvest or beyond is used as a base. It is sought after for its sweet fragrance and roasted aroma, but depending on the base, *Hojicha* can be different in both taste and flavor. *Hojicha* made from stems tends to be light-bodied with a sweet fragrance, whereas a leaf-based *Hojicha* is stronger on the whole, with a prominent astringency. In Japan, it is mainly drunk after meals. If you happen to have Sencha that has gone old, do not despair as you can turn it into Hojicha by carefully roasting it in a skillet (however, make sure to use one that does not have any food odor). This will not only revive the tea, but the pleasant roasted aroma will also enliven your surroundings.

茶茶/棒茶 | Kukicha / Bocha

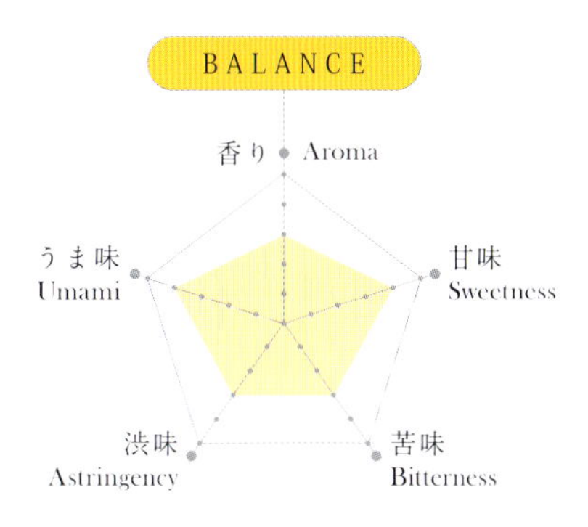

美味しい淹れ方
How to steep Kukicha and Bocha ? → page 68

茎茶または棒茶は、主に生産者の作った荒茶を製茶問屋が仕上げ加工をする際に取り除いた茎の部分で作られます。そのため、生産量の統計はありません。青々とした草原のような香りが特徴です。渋味が少なく、やや熱めのお湯で淹れても美味しいため食事に合わせたり、肌寒い時期などに飲むのもおすすめです。茎だけを焙煎したものを「茎ほうじ茶」、「棒ほうじ茶」と呼びます。

Kukicha, also called *Bocha*, is made from stems and stalks that are sifted from *Aracha* by the wholesaler during the refining process of steamed green teas. Because it is not made for the sake of itself, there are no statistics on the production volumes. The vegetal notes resembling grass covered plains are characteristic to *Kukicha*. Being weak in astringency, it can be steeped at slightly high temperatures as well. This is an advantage when you are pairing it with food or would like to warm yourself up on a cold winter day. When *Hojicha* is made from *Kukicha*, it is usually called *Bo Hojicha* or *Kuki Hojicha*.

粉茶 | Konacha

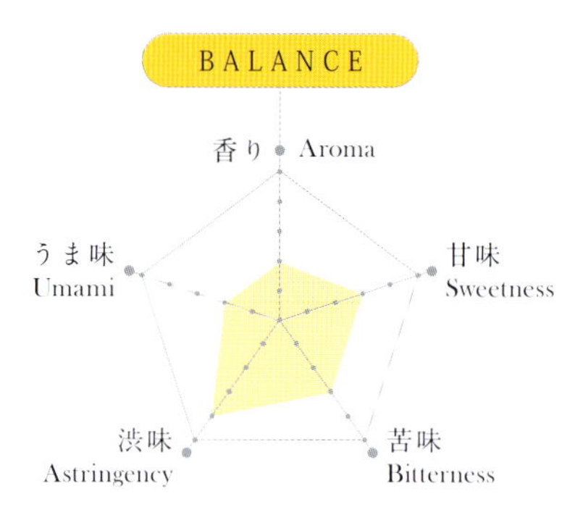

美味しい淹れ方
How to steep Konacha ? → page 84

粉茶は日本の寿司店の定番のお茶です。生魚を使った料理などを食べた後に、熱湯で淹れた粉茶を飲み、口中をさっぱりとさせたり、カテキン類の殺菌力による食中毒の予防を期待します。粉茶も茎茶と同じく、荒茶の仕上げ加工時に選別されたお茶です。ふるい分けと風選によって選別した重量の軽い部分が粉茶になります。リーズナブルなティーバッグの中身にも使用されます。粉茶の文字から「粉末茶」と混同されることもありますが、粉末茶は茶を粉末状に加工したもので、粉茶とは別のお茶です。

Konacha is a fixture at Japanese sushi restaurants, and goes well with dishes that include raw fish for two reasons. Steeped in boiling hot water, it clears the palate and refreshes us, but the catechins may also lower the risk of food poisoning. Just like *Kukicha*, *Konacha* is also something that comes about as a result of the sifting process during tea refining. The light parts (green tea dust or fannings) that are sifted and sorted ends up as *Konacha* , and due to its comparably low price it often ends up as a teabag tea. It is often confused with powdered green tea (*Funmatsucha*) but powdered green tea is a ground type of tea, which makes it very different from *Konacha*.

釜炒り茶 | Kamairicha (Pan-fired green tea)

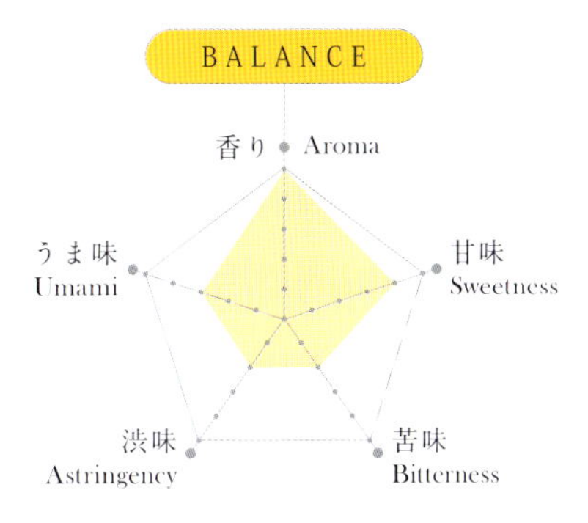

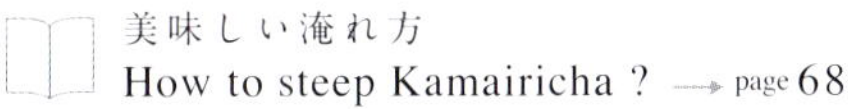
美味しい淹れ方
How to steep Kamairicha ? → page 68

ほとんどの日本茶は生葉を蒸すことによって作られ、釜で炒って作るのは一般的に中国式と言われています。日本で作られている炒り緑茶ももともと中国から渡来しましたが、煎茶よりも歴史が長いことはあまり知られていません。現在では宮崎県や佐賀県など主に九州で作られていますが、中国の釜炒り茶とはまた違う味わいを醸し出すお茶です。日本茶らしいうま味がありながら渋味が少なく、爽やかな釜香で気分がすっきりします。玉緑茶とも呼ばれますが、煎茶の最後の揉む工程を省いた蒸し製緑茶の玉緑茶とは名前が似ていても、製法も味も異なります。

Almost all teas in Japan are made by steaming the fresh leaves, whereas pan-firing is usually referred to as the Chinese way of processing green tea. Pan-fired teas in Japan also have their origin in China, and the fact that they have a longer history than *Sencha* is not well known. As of today, Japanese pan-fired teas are mainly produced in the prefectures of Miyazaki and Saga on the island of Kyushu in the Southern part of Japan, and the taste is different from its Chinese relatives. You can sense a very Japanese touch of *umami*, but it has less astringency than *Sencha* and the distinct sweet roasted aroma refreshes both the palate and the mind. Pan-fired Japanese teas are often called *Tamaryokucha*, but should not be confused with steamed *Tamaryokucha*, which is produced in a way similar to that of *Sencha* but with the last rolling step omitted.

碾茶 | Tencha

碾茶の存在を産地以外で知る消費者は少ないかもしれません。碾茶は抹茶の原料となる茶で碾茶として流通することは稀です。碾茶を石臼などで碾(ひ)いて粉末化したものが抹茶となります。碾茶は、玉露と同様に被覆して栽培されますが、製造方法が大きく異なります。水蒸気を利用した加熱の後、速やかに揉まずに高温で乾燥させ、茶葉から茎だけでなく葉脈など硬い部分を取り除くことで、魚のうろこ状の形のお茶が出来上がります。高品質な碾茶は手摘みの一番茶ですが、二番茶以降の機械摘みなど全く違う条件で製造されたものもあり、これらは菓子類などに使われる加工用抹茶の原料となります。良質な抹茶を知るためにも、機会があれば高品質な碾茶に触れたいものです。

Apart from those who live in regions where they actually produce it, probably few consumers know about *Tencha*. *Tencha* is ground, usually in a stone mill, and turned into the powdered bright green tea that we call *Matcha*. It is therefore rare to find in its original form. *Tencha* is shaded before harvest just like *Gyokuro*, but processed in a completely different way. After steaming, it is dried fast in high temperature without rolling, after which not only the stems but also all the other hard parts, like the veins, are separated from the leaves, resulting in a flaky tea resembling fish scales in shape. High quality *Tencha* is plucked by hand in the first harvest season, but mechanically harvested *Tencha* from later seasons is processed under very different circumstances and can also be found on the market. This becomes what is often referred to as culinary *Matcha*, usually used for sweets and pastries. Coming in contact with high quality *Tencha* will benefit anyone who wants to become able to distinguish good *Matcha* from lower grades.

抹茶 | Matcha

美味しい点て方
How to whisk Matcha ? → page 82

抹茶は現在日本で作られている最も歴史のある茶種の一つです。近年はお菓子やラテ用の素材として知名度が国内外でも上がってきましたが、本来は茶道も含め、抹茶単体をお湯で溶き混ぜて飲むお茶です。栽培及び製造の方法は煎茶などとだいぶ異なり、碾茶（28頁参照）を製造して後に茶臼で挽いたものが抹茶として認められます。良質な抹茶を点てると鮮やかな濃い緑色とふくよかな香りがあり、豊かなうま味と甘味、渋味が素晴らしいハーモニーを奏でるようです。上質な抹茶本来の味わいを存分に堪能出来る濃茶、カジュアルな薄茶、ともに是非、茶筅を用いて抹茶茶碗で点て、一服を楽しんでください。

Matcha, one of the oldest types of tea in Japan, has in recent years gained popularity as an ingredient in confections, even overseas. Originally however, it was dissolved in hot water and drunk at occasions like the Japanese tea ceremony. *Tencha* (see p.28) ground in a tea mill is considered to be proper *Matcha*, so both cultivation and processing differ greatly from *Sencha*. When whisked, high quality *Matcha* yields a vivid deep green color, and the rich aroma, in concert with a strong *umami* and sweetness balanced with a pleasant astringency, make up a perfect harmony. High quality *Matcha* can be enjoyed to the maximum as "*Koicha*" (thick *Matcha*, where more powder and less water is used) or more casually as "*Usucha*" (light *Matcha*, where less powder is used). Any tea lover should definitely try both. Using a traditional tea whisk (*chasen*), and enjoying the tea from a genuine Japanese bowl will give the best experience.

国産紅茶 | Japanese black tea

淹れ方	Steeping guide
茶葉の量　3g	Tea leaves　3g
熱湯　250cc	Water amount　250ml
時間　約4分	Steeping time　About 4 minutes

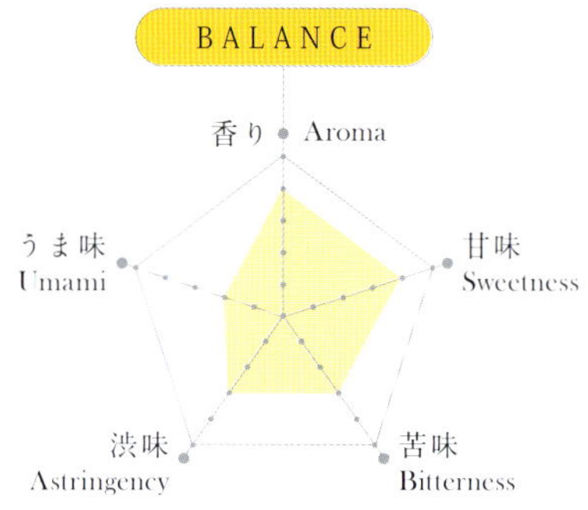

紅茶が日本で作られていたことはあまり知られていませんが、お茶の輸出が盛んだった明治時代から昭和中期までは本格的に紅茶製造に取り組んでいた産地が幾つもありました。インドから紅茶作りに向く品種のチャの種を持ち込み、日本国内で栽培を行っていました。近年注目された「べにふうき」や「べにひかり」などはかつて栽培された紅茶用品種に由来します。貿易の自由化など他の生産国に対する競争力低下や、緑茶の国内需要の高まりから、生産量が非常に少なくなりましたが、最近、日本産の紅茶が改めて注目されています。スリランカやインドなど、伝統的な生産国の紅茶とは香味が異なる「和紅茶」は、渋味が少なく甘味を感じ、砂糖や牛乳などを加えずにストレートでも美味しく味わえます。

Few associate Japan with black tea, but from the Meiji period (1968-1912) until the middle of the Showa period (1926-1989) some of the tea producing regions in Japan where engaged in large scale black tea production, mainly for export. Seeds from Indian tea plants suitable for the processing of black tea were even brought to Japan, and the cultivation of these led to new crossbreeds. *Benifuki* and *Benihikari*, two cultivars that have received attention in recent years, are examples of offspring that share genetic material with the black tea cultivars once widely grown. After the liberalization of trade with black tea in Japan, which coincided with a growing domestic demand for green tea, black tea production in Japan became less competitive, and as a result it was set on a path that almost led to its extinction. However, in recent years Japanese black tea, or "*Wakocha*" as it is often called, has seen a revival, and the mild sweet taste with almost no astringency makes it different from tea from classic black tea countries, such as Sri Lanka. It tasted good as it is, without any added milk or sugar.

碁石茶 | Goishicha

淹れ方	Steeping guide
茶葉の量　2g	Tea leaves　2g
熱湯　200cc	Water amount　200ml
時間　1分	Steeping time　1 minutes

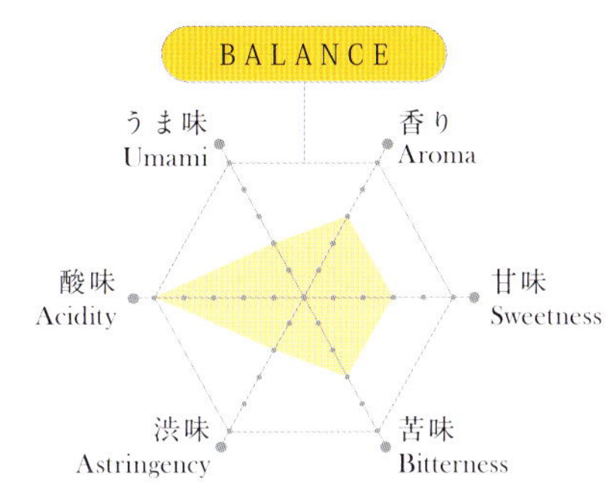

黒いブロックのような見た目から、お茶には思えないかもしれませんが、古くから高知県で作られている独特な後発酵茶です。生葉を蒸してから7〜10日の間、カビつけをして発酵させた後、桶につけて乳酸菌を利用した発酵を行います。2〜3週間後に切り分けて取り出し、さらに小さくブロック状に切ってから天日干しにします。出来上がったお茶は酸味が強く他の茶種と香味が全く異なります。昔はおかゆなどを作るための素材でしたが、現在では珍しいお茶として重宝され、あえてストレートで飲む人が多いようです。四国地方には、碁石茶以外にも、愛媛県の石鎚山の黒茶や徳島県の阿波番茶などの後発酵茶がありますが、どれも生産量が少なく非常に珍しいお茶です。

Due to its unique and rather odd appearance, it is perhaps hard to believe that this is tea at all, but this fermented rarity has been produced in Kochi prefecture on the island of Shikoku since long back. After steaming, the leaves are stacked and left to mold for 7 to 10 days, then packed in a barrel where it gets fermented with lactobacillus. After two weeks, the tea is taken out, cut into small blocks and dried in the sun. The result is a tea unlike most others, with its prominent sourness and dark liquor. In the old days, this tea was used as an ingredient in rice porridge, but nowadays it is treasured as a rare tea, and therefore even drunk as a pure tea without any additives to enjoy its originality fully. Apart from *Goishicha*, on the island of Shikoku we also find similar teas, such as Ishizuchi San *Kurocha* from Ehime prefecture and *Awabancha* from Tokushima prefecture. All of them are only produced in small quantities and are very rare to come across today.

可搬型摘採機による摘採 / Tea picking with a mechanical harvester

日本茶の味のヒミツ
The secret behind the flavor of Japanese tea

生葉
Fresh leaves

蒸熱後
After steaming

葉打後
After rattling the leaves

粗揉後
After primary/rough rolling

煎茶（荒茶）が出来るまで

日本では冬に休眠した茶の樹が春になると目覚めたように、栄養を豊富に含んだ柔らかな新芽を伸ばします。この新芽がその年の一番茶になります。新芽は特に寒さに弱いので、茶農家は防寒・防霜の対策を行いながら成長を見守ります。新しい葉を５枚ほど展開した頃、一芯二葉～三葉を目安に摘みとりが行われます。日本では機械による刈り取りが主ですが、最も高品質な茶は手摘みが基本になっています。摘みとった生葉は萎(しお)れないように涼しいところで保管し、製造に必要な分量がまとまったら速やかに蒸します。紅茶や烏龍茶など釜炒り製の茶の工程にある「萎凋(いちょう)」が日本の蒸し製緑茶にはありません。そのため自然な香味が保たれたまま、次の乾燥工程に入ります。まずは表面の水分を飛ばすため、熱風を加えながら葉を攪拌(かくはん)します。ある程度乾いたら、次に圧を加えて揉みながらさらに乾かしていきます。この際に、出来るだけ茶葉を破砕させずに乾燥度を高めていくのが製茶の基本です。蒸してから荒茶が出来上がるまで、約５時間です。茶葉の状態が早いペースで変わりますので、生産者は限られた短い時間の中で茶葉の状態などを判断しなければなりません。茶葉は直接、機械や手で揉むのではなく、お茶をお茶で揉むようにします。均一な乾燥には、原料となる茶葉を揃えることが重要になります。そのため高品質な茶作りは、念入りな茶園の管理と適切な摘み方が必須です。最終工程の乾燥機に入る前、精揉(せいじゅう)という工程で煎茶は針のような形になります。生葉は蒸すことによって柔らかくなり、元来、形の異なる葉と茎を同じような針状の形へと揃えて縒(よ)っていくことが出来ます。椿の仲間であるチャは葉の表面に艶があり、一芯二葉～三葉で摘みとりがされれば、艶のある面が表に出やすくなります。これが高品質の日本茶が、濃緑色で宝石のような光沢を持つ理由のひとつです。

蒸す工程 | The steaming process

精揉の工程 | Final rolling

揉捻後
After rolling/kneading

中揉後
After intermediary rolling

精揉後
After fine rolling

乾燥後
After drying

Different stages of Sencha (Aracha) manufacturing

After winter dormancy the Japanese tea plants regain their liveliness in spring, and new soft buds rich in nutrition also come alive. These young buds will grow and eventually turn into *Ichibancha*, the first harvest of the year. Since they are very sensitive to cold, the farmers will take great care to protect the young buds from any frost or cold damages as they grow. When up to five new leaves have grown out, the time is ripe for harvesting, and usually the bud is picked together with the two, or sometimes three, leaves on the side. In Japan it is common to use different types of mechanical harvesters, but the best quality teas are plucked by hand. The plucked leaves are kept in a cool, dark place to prevent any withering, and when there are enough leaves for one batch they will be steamed as soon as possible. There is no withering process for Japanese green teas, which makes it different from black tea, oolong tea and many pan-fired green teas. As a consequence, the natural flavors remain intact as the leaves are sent to the next stage. First, the leaves are rattled in order to get rid of the moisture on the surface. When the surface dries, pressure is added to extract the moisture of the inner part of the leaves. To turn the leaves into a dry product without crushing them is a fundamental part of tea making. It takes about 5 hours from steaming to finished *Aracha*, and since the condition of the leaves changes fast, the tea producer has to take many difficult decisions over only a short time. Tea is not rolled directly by hand or by machine, but the leaves are rolled against each other. In order to dry the leaves evenly, it is important to pick tea of as similar shape and size as possible. In other words, high quality tea requires that great care is taken of the plantation, but also that the tea is carefully plucked. Before the final drying process, the leaves go through a "fine rolling" stage where they are stretched out to the characteristic needle like shape. Stems and leaves originally differ in shape, but since steaming makes the leaves soft and malleable, they eventually end up looking the same as they are rolled and dried together. Being a Camellia, the tea leaf has a glossy side, and this side tend to make up the surface of the dried leaves when "two leaves and a bud" or "three leaves and a bud" are picked carefully. This is the reason why high quality Japanese tea tend to have the deep green color and glossy leaves that almost makes it resemble a precious stone.

製茶 — 荒茶と仕上げ茶
Tea refining - Aracha and Shiagecha

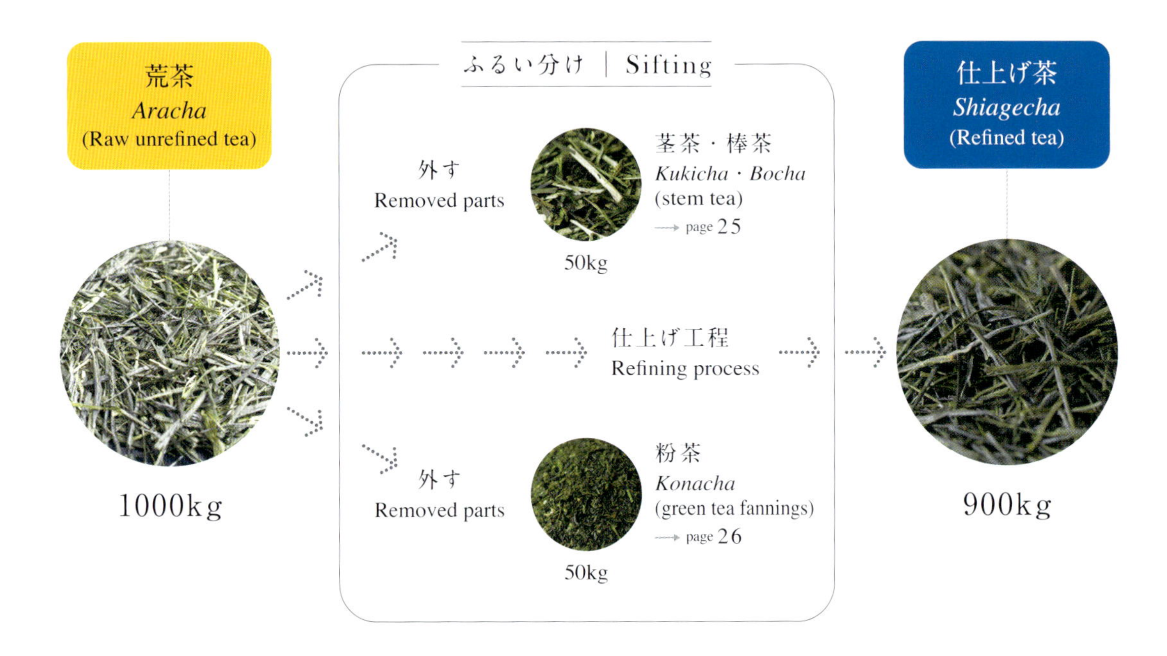

日本茶を知る上で、「仕上げ」とは何か、そしてなぜ行われているのかを理解することが大事なポイントです。日本茶は茶農家である生産者が「荒茶」を作ります。荒茶は余分な茎や製茶時に出来た粉などが混然一体となったもので、販売用茶葉の原料です。仕上げによって荒茶の中にある雑味の元となる粉や茎、大型の葉や異物を取り除き乾燥度を高めます。荒茶の水分含有量は約5%、仕上げ後は3%以下となります。乾燥度が上がるとお茶の香りや味わいは、よりひき立ちます。この乾燥工程を「火入れ」と呼びます。これらの工程を終えたものを「仕上げ茶」と呼びます。そして仕上げの役割を担うのが製茶問屋です。品質を揃えるためのブレンドも含めて、日本茶において重要かつ欠かせない仕事です。

In order to understand the world of Japanese tea, it is crucial to understand what tea refining is and why it is carried out. Japanese tea farmers produce something called *Aracha* (often translated as "raw tea" or "unrefined tea"), which contains both an excessive amount of stems and also dust-like leaf particles that appear during the manufacturing process. At this stage the tea is still considered a raw material rather than a final product. Parts that give the tea a harsh taste such as fannings (tea dust) are sifted and stems and large leaves, as well as anything that does not belong in the tea is removed. Another important step is to reduce the moisture level. *Aracha* still has about 5% moisture whereas refined tea, *Shiagecha*, has 3% or less. This drying process is called *hiire* in Japanese (written with the Chinese characters "fire" and "enter", often translated as "firing" or "finish-firing") and reducing the moisture will enhance both the aroma and taste. Tea refining in Japan is mainly done by wholesalers, and in order to maintain a high quality this process is indispensable.

手揉みについて
Hand made Japanese tea

日本茶はほぼ全て機械で作られています。手揉み茶は、商品を作るというよりも伝統技術の伝承が第一義の目的です。手で煎茶を上手に揉むには、体力や根気、そして長年の経験と洗練された技術が必要になります。機械でお茶を作る時（32頁参照）と同じように、最初は表面の水分を飛ばし、その後は徐々に力を加えてお茶でお茶を揉むようにしながら乾燥させていきます。これは「焙炉（ほいろ）」という表面に和紙を敷いた木製の作業台で行われます。焙炉の下には熱源が置いてあり、かつては炭を使っていましたが、現在はガスバーナーを用いるのが一般的です。一人の手揉み師が一番茶を揉んだ場合、２kgの生葉が５時間の重労働を経て、約400gの乾燥した茶葉になります。機械化が進む前は、日本茶を作るのにどれほど苦労と多くの人力が必要であったか想像出来るでしょう。

Almost all teas in Japan are mechanically processed (see p.32), and hand rolled tea is mainly made by tea producers to pass on the traditional craft rather than for commercial purposes. Making hand made *Sencha* is extremely labor intensive and requires both bodily strength, mental persistence and, needless to say, a lot of skill and many years of experience. Just like when tea is processed mechanically, at first the moisture of the surface is removed by rattling the leaves. Later on, pressure is added and the leaves rolled and kneaded against each other to extract the moisture of the inner part of the leaves. This is done on a *washi* (Japanese traditional paper) covered wooden work table called *hoiro*. The heat source used to be charcoal, but nowadays it is more common to use a gas burner. If a tea master makes one batch of *Ichibancha*, 2kg of fresh leaves would after 5 hours of hard labor end up as 400g of dried leaves. One can imagine how hard the tea producers must have worked in the past before tea processing was mechanized.

合組
Gogumi (Blending)

「色は静岡、香りは宇治よ、味は狭山でとどめさす」。いつの頃からか埼玉でこのようなPRがされるようになりました。これはもちろん単純化された表現で、実際にそれぞれの産地の特徴を正確に説明しているわけではありませんが、近代以降、日本茶がブレンドされて作られてきたことを象徴しているように思います。その背景には、日本茶の流通の歴史があります。生産者は、半完成品である荒茶（34頁参照）を作ります。荒茶を完成品に仕上げるのは、設備を持つ製茶問屋の仕事です。複数の生産者から荒茶を仕入れた製茶問屋は、仕上げだけでなく、品質の均一化と価格の安定を目指して合組（ブレンド）を行います。この作業により、個々の生産者の顔は見えないものになりますが、茶問屋のそれぞれの個性が現れます。問屋が多く集まるのは宇治と静岡ですが、この二か所は産地だけでなく、他の生産地の荒茶を仕上げや合組を行い、消費地に送る集散地でもあります。

"Shizuoka for the color, Kyoto for the aroma, and Sayama for the taste". This is a known phrase in the tea industry, and has been used for quite some time to promote tea from Sayama. The descriptions are of course simplifications, and the characteristics of the mentioned regions are not described accurately. That said, I believe that this saying very well expresses how Japanese tea is seen as something that is blended, a practice that has dominated the industry ever since modernization began. The reason for this can be found in the history of tea distribution. In Japan, a tea producer does not make a final product but something called *Aracha*(see p.34) and tea wholesalers, who have the equipment for refining, turn it into a finished product. However, as the wholesalers purchase *Aracha* from many different producers, they do not only refine the tea, but also blend it to even out the quality and stabilize the price. From the perspective of a tea producer, this makes the tea lose its personal touch, but on the other hand the tea refiners each have their own style which give the tea its unique characteristics. Tea wholesalers/refiners are concentrated in Uji and Shizuoka, two regions that apart from being tea growing areas also act as big refining and distribution centers, were most Japanese tea pass through before ending up on the shelves of tea shops etc.

日本茶の審査
Japanese Tea Evaluation

日本茶の審査

茶の「審査」や「鑑定」と言うと、産地や生産者、品種などの違いなどを探ることをイメージする人が多いかもしれません。しかし、実際の日本茶の審査で最も重要なのは、対象となるお茶の欠点の確認と品質の価値づけをすることです。欠点とは、栽培または製造のどこかの段階で生じた異臭や雑味などのことです。特に食品としては大きな問題がある煙臭や油臭、薬臭、そして製造不良の有無を確認します。このような審査は、製茶問屋などが主に担います。ここで品質確認がされているため、不良茶はほとんど市場には出回らないのです。品質を客観的に判断して、生産者にフィードバックするのも審査の目的です。製造時の問題点を伝えて次の製造改良に役立つようにします。

日本茶の審査は、水色と茶葉の色がわかりやすいように白い磁器の茶碗を用います。そこに荒茶のサンプルを3gずつ入れ、欠点を感知しやすいように熱湯を注ぎます。網匙で茶葉を掬って香りを嗅ぎ、その後、スプーンで味を確認します。この作業は、審査して品質を判断するだけでなく、茶葉の特色を掴み、仕上げ方と合組をどのように行うのかをなども決めていきます。

Quality evaluation of Japanese Tea

Many people probably imagine that tea evaluation or grading is about being able to distinguish the difference between tea from a number of regions, producers and cultivars. However, the main point of quality evaluation is to detect any defects and also to determine the value of tea. Examples of defects could be off-flavors and harsh tastes that appear either during cultivation or processing. Especially smokiness, oily smell or smell of remaining pesticides would be a grave problem for any food product or beverage, and tea is also checked for any manufacturing failure. This type of tea evaluation is mainly carried out by tea wholesalers/refiners so tea with the above mentioned flaws does in principle not appear on the market. Another point here is to give the producers feedback on their tea based on objective evaluation. Any defects or problems are reported and this information helps the producers improve their processing skills.

For evaluation of Japanese tea, small bowls made of white porcelain is used, making it easier to assess the color of both the infused leaves and the liquor. 3g of *Aracha* is used, and to make any flaws or defects as easily perceived as possible, boiling hot water is used. A mesh strainer spoon is used for assessing the aroma (pictured) and thereafter a teaspoon is used for tasting. Not only the quality is checked, but the taster also takes notes on the characteristics and possible ways o refining and blending.

シングルオリジン
Single Estate Japanese tea

シングルオリジンとは

ほとんどの日本茶はブレンドがされているのに対して、最近、いわゆるシングルオリジンの日本茶が話題になっています。シングルオリジンとは、特定の茶園で摘みとった原葉で出来た荒茶を合組（36頁参照）せずに仕上げたお茶のことですが、ワインのシャトーと同じように品種や産地によって生まれる個性を楽しめます。そもそも単一品種のシングルオリジンは、品種導入が進んできた戦後に初めて可能になりましたが、ブレンドせずに単品で仕上げる動きは2000年以降に加速しました。ここでは日本の代表的な茶の品種を紹介します。

What is Single Estate Japanese tea?

Although almost all Japanese teas are blends, single estate Japanese tea is getting increasingly more attention lately. Single estate Japanese tea is, as the name implies, tea leaves picked from one single tea garden processed into *Aracha*, and then refined without blending. By drinking single estate teas, we are able to enjoy the taste and flavor peculiar to the chosen cultivar or the terroir just like single vineyard wine. Cultivars were introduced on a large scale only after the second world war, and the single estate movement gained momentum as late as in the beginning of the 21st century. Here I will introduce some of the cultivars that are representative of Japan.

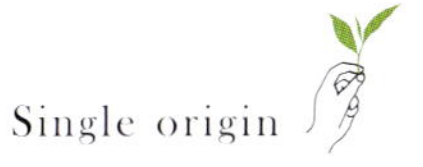

品種茶の系譜図
Lineage of selected Japanese tea cultivars

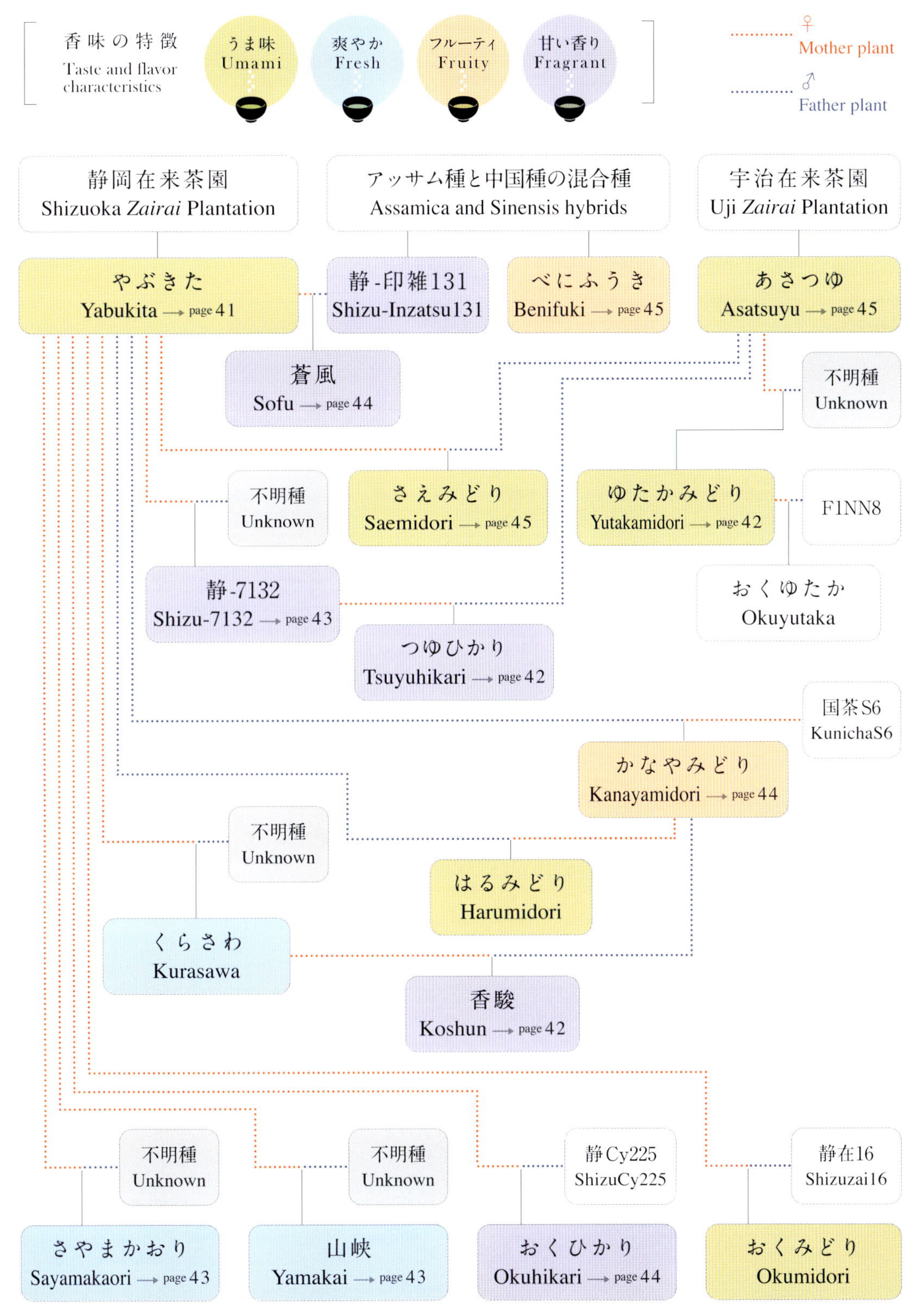

在来種 | Zairai (Seed propagation)

在来種／岐阜県春日の茶園
Zairai plantation in Kasuga, Gifu Prefecture

在来種の茶葉―種子発芽
Zairai ― Propagation from seeds

クローン栽培の茶葉―挿し木
Cultivar (clones)― Propagation from cuttings

チャは自らの花粉では実を結ばない植物のため、品種茶は全て、挿し木で植えたクローンです。それゆえに品種茶の園は概ね新芽が均一に伸びていきます。在来茶園は自然交配によって出来たチャの種を蒔いて作った園地で、茶樹の一本一本が異なり、新芽の色や形、伸び方も均一ではありません。姿だけでなく、香味もそれぞれ違うので、ある意味で自然のブレンドとも言えます。香味は品種茶のようにはっきりとした特徴がありませんが、お茶の持つ野趣が最も際立っているのが在来の魅力です。かつて、茶園は全て在来種でしたが、現在は全体の約2%ほどしか残っていません。

The tea plant does not self pollinate, so tea cultivars are all clones propagated by cuttings, and as such they grow almost evenly into perfect hedges. On the other hand, in a so called *zairai* plantation all the tea plants are propagated by seeds that are the result of natural cross pollination. As a consequence, individual plants grow at a different pace, and the color and the shape of the leaves are also different. Not only the appearance, but every single plant also has its own distinct taste and flavor making *zairai* tea something of a natural blend. The taste and aroma is not as clear and refined as cultivar tea but if wilderness or nature can be sensed in green tea in general, this attribute is by no doubt most strongly accentuated in *zairai* tea. Before the introduction of clonal cultivars all tea plants used to be propagated in this way, but nowadays *zairai* covers only about 2% of all tea plantations. *Zairai* is often translated as "Japanese native", which is misleading since tea is not native to Japan, but thought to have been brought from China by Buddhist monks.

やぶきた | Yabukita

日本茶にはうま味と甘味、渋味、苦味の４つの要素がありますが、そのバランスが最も良く取れている品種は間違いなく「やぶきた」です。外観も綺麗で、高品質の「やぶきた」は日本の山を思わせる馥郁たる香りが立ちます。1908年に杉山彦三郎によって静岡の在来茶園から選抜され、現在、日本の茶園の75%ほどを占め、日本茶の軸とも言える存在です。産地や栽培方法、製法などによって違う姿を見せるので、飲めば飲むほどその奥深さに気付かされます。

Yabukita is, without doubt, the cultivar that has the best balance of the four taste elements that make up Japanese tea; *umami*, sweetness, astringency and bitterness. It also has a beautiful appearance, and good quality Yabukita will produce a sweet aroma that carries one's thoughts to Japan's mist clad mountainous regions. It was singled out from a *zairai* plantation in Shizuoka by Hikosaburo Sugiyama in 1908, and as it covers a total of 75% of all the tea plantations in Japan, it can in a way be said to serve as a standard for Japanese tea.
Depending on the region and processing method the taste and flavor will end up different, so by trying different Yabukita teas, one is struck by the depth of this truly unique cultivar.

[品種登録・年度] 静岡県茶業研究センター（1953年）
※1908年に杉山彦三郎により選抜
[Institution and registration year]
Shizuoka Prefecture Tea Research Center (1953)
*Selected by Hikosaburo Sugiyama in 1908

築地東頭

静岡の山奥、標高800mの土地「東頭」に、徒歩でなければ辿り着けない美しい茶園が隠れています。傾斜面で寒暖差が激しいこの天空の茶園のお茶は、年に一回、丁寧に手で摘み採ります。品種は主に「やぶきた」ですが、この独特な環境に育まれたやぶきたは他の土地のお茶にはない深みある香りが漂います。日本茶の理想を求め、あえてこの厳しい山間地を開拓した生産者が築地勝美氏。氏の努力と技術によってすばらしいシングルオリジンの日本茶が誕生しました。

Tsukijitobetto

800m above sea level deep in the mountains of Shizuoka, a stunningly beautiful tea garden is hidden that can only be reached by foot. Due to its harsh environment, with steep slopes and a great temperature difference, tea is only picked once a year, carefully by hand. The cultivar grown is Yabukita, but due to its peculiar environment the tea has a deep aroma that can only be found there. In the pursuit of the ideal *Sencha*, Katsumi Tsukiji, who cleared and cultivated this area, called Tobetto, managed to create a wonderful example of a single estate Japanese tea.

香駿 | Koshun

目を瞑って「香駿」を飲むと、まるで花畑の中にいるかのように感じます。その香りは、多くの人を虜にします。日本茶の中ではフローラルな香りが豊かで、余韻も長く残ります。苦味が若干強く、ビールに例えるならエールのような存在に当たります。

Close your eyes while sipping on a Koshun *Sencha* and you might think that you are standing in the middle of a garden in full bloom. More and more tea drinkers are charmed by this floral aroma, but Koshun also has a pleasant aftertaste that lasts surprisingly long. The fruity notes are balanced by a comparably strong bitterness and in the world of beer you would find the equivalent in an IPA.

[品種登録／年度] 静岡県茶業研究センター／2000年
[Institution and registration year]
Shizuoka Prefecture Tea Research Center / 2000

ゆたかみどり | Yutakamidori

「やぶきた」に次いで栽培面積が2番目に多い品種で、そのほとんどが鹿児島で栽培されています。静岡生まれの品種で、やや苦味を持ちますが、多くが深蒸し煎茶になるため、基本的には刺激が少なく飲みやすいお茶です。根菜類のような甘味と香りを感じるとも言われます。

After Yabukita, Yutakamidori covers the second largest area of all Japanese cultivars. Although bred in Shizuoka, most of it is grown in Kagoshima prefecture. It has a slightly strong bitterness, but since most of it is processed into *Fukamushi Sencha*, Yutakamidori teas tend to have mild and smooth taste with a sweetness reminiscent of root crops.

[品種登録／年度] 農研機構果樹茶業研究部門 金谷／1966年
[Institution and registration year]
National Institute of Fruit Tree and Tea Science, Kanaya Japan / 1966

つゆひかり | Tsuyuhikari

「つゆひかり」は、桜葉のような香りが特徴的な「静-7132」と根菜類を思わせる香りがある「あさつゆ」の間で生まれた新しい品種です。それぞれの特徴をバランスよく持ちながら甘味が比較的に強く、渋味が苦手な人でも楽しみやすい品種とされます。

Tsuyuhikari is a crossbreed between Shizu-7132 and Asatusyu. The former has notes of cherry blossom, whereas the latter has more prominent vegetal notes. Apart from a good balance between the attributes of its parents, Tsuyuhikari is also comparably sweet and is therefore treasured as a tea that can be drunk by people who prefer tea weak in astringency.

[品種登録／年度] 静岡県茶業研究センター／2003年
[Institution and registration year]
Shizuoka Prefecture Tea Research Center / 2003

甘い香り
Fragrant

静-7132 | Shizu-7132

日本の春といえば桜が有名ですが、この品種は香料が入っていないのに桜の香りが漂います。時期や場所などを問わず、いつでもどこでも日本の春を彷彿とさせる、とても魅力的な品種です。正式登録されなかったことから名前は系統番号のままになっています。

The Japanese spring is famous for its cherry blossom, or Sakura in Japanese. Having an aroma resembling Sakura, without any added flavoring, this wonderful cultivar always offers a delightful reminder of the beauty of Japanese spring, regardless of time and place. As it was never registered, it never got a proper name but still goes by the number it was given at the research stage.

[品種登録 / 年度] 静岡県茶業研究センター（未登録）
[Institution and registration year]
Shizuoka Prefecture Tea Research Center (never registered)

フルーティ
Fruity

山峡 | Yamakai

氷を使った冷水で濃く淹れるとメロンのような味がする珍しい品種です。リッチでありながら同時にとても爽やかに感じます。豊かなうま味からかつて「天然玉露」と呼ばれたこともあります。高品質の「山峡」は熱めのお湯で、香りを立てて淹れても美味しく味わえます。

When making a thick rich infusion using iced water and a large amount of leaves, not only a strong *umami* but also the melon notes peculiar to this rare cultivar will become stronger. Due to its rich t*aste*, it also goes by the nickname "natural *Gyokuro*", but unlike *Gyokuro*, high quality Yamakai can be steeped even in hot water.

[品種登録 / 年度] 静岡県茶業研究センター / 1967年
[Institution and registration year]
Shizuoka Prefecture Tea Research Center / 1967

爽やか
Fresh

さやまかおり | Sayamakaori

茶園では新芽が揃って針のように伸びる姿がユニークです。うま味が少なく、ワインの用語で例えると喉越しはライトボディー。爽やかな渋味と品種の香りが森林を想わせて、気分がすっきりします。品種ならでは特有の甘さがあるとても面白い品種です。

The way the leaves grow evenly almost like needles at the plantation makes this cultivar look unique. Comparably weak in *umami*, it has a mouthfeel that could be described as light-bodied, and the refreshing astringency and forest-like aroma invigorates the mind. The characteristic sweetness make this cultivar both unique and interesting.

[品種登録 / 年度] 埼玉県茶業研究所 / 1971年
[Institution and registration year]
Saitama Prefecture Tea Research Center / 1971

かなやみどり | Kanayamidori

ミルキーな香りが特徴とされる品種ですが、栽培や製造の条件によって印象はかなり変化し、煎茶にすると柑橘類を思わせるフルーティーなお茶にもなります。冬から春への季節が移り変わる頃に飲むと春がひと足先に来たような気がします。

This cultivar is known for having a milky flavor but depending on the growing conditions and processing method, it changes considerably, sometimes having fruity notes resembling citrus. If drunk at the time when winter turns into spring, it certainly gives good spring vibes.

[品種登録/年度] 農研機構果樹茶業研究部門 金谷/1970年
[Institution and registration year]
National Institute of Fruit Tree and Tea Science, Kanaya Japan / 1970

蒼風 | Sofu

インドのマニプールから持ち込んだ種子に由来する遺伝子を持つためか、ダージリンのようなマスカットを思わせる香りが特徴です。日本茶らしい豊かなうま味とフラワリーな風味を持っています。また、お茶の葉を少なめにして紅茶のように熱めのお湯で淹れても美味しく味わえる希少な品種です。

Unique muscatel notes, perhaps a result of its partly Indian genome, and the flowery aroma are the main characteristics of this cultivar, but it also has typical Japanese attributes such as a strong *umami*. It tastes good even when steeped like black tea, in hot water and with slightly fewer leaves.

[品種登録/年度] 農研機構果樹茶業研究部門 金谷/2002年
[Institution and registration year]
National Institute of Fruit Tree and Tea Science, Kanaya Japan / 2002

おくひかり | Okuhikari

「やぶきた」より力強く、加えて甘い香りが立ち上がる、硬質で芯のある香味が特徴です。長年緑茶を飲みなれている人にはたまらない香りです。物々しさが全くなく、侘び寂びを思わせる、日本らしい美味しさが感じられる品種です。

A little stronger than *Yabukita* and with a sweet aroma, this cultivar shows both steadiness and persistence in taste as well as flavor. As such, it is cherished by many devoted green tea drinkers. Although it lacks grandioseness, it has a subtle beauty, making it a gustatory expression of the Japanese aesthetic ideal *wabi sabi*.

[品種登録/年度] 静岡県茶業研究センター/1987年
[Institution and registration year]
Shizuoka Prefecture Tea Research Center / 1987

さえみどり | Saemidori

ある意味で「やぶきた」に似ていますが、渋味と苦味が少なく、初心者にとっても飲みやすいお茶です。太陽の光が強くてもあまり苦くならないので、うま味と甘味を中心に味わえる品種です。鹿児島など平坦な産地で普及しています。

In a sense Saemidori is similar to Yabukita, but with less bitterness and astringency it makes a good first green tea for beginners. This cultivar, which is enjoyed mainly for its sweetness and *umami*, does not become too bitter even when exposed to strong sunlight, which is one reason for its popularity in flat tea growing regions like Kagoshima.

[品種登録／年度] 農研機構果樹茶業研究部門 枕崎1990年
[Institution and registration year]
National Institute of Fruit Tree and Tea Science, Makurazaki Japan / 1990

あさつゆ | Asatsuyu

京都の在来茶園（40頁参照）から選抜された品種ですが、青野菜や枝豆を思わせるような香りが「緑」を感じさせます。「やぶきた」がシャープな面があるのに対し、「あさつゆ」は丸みのあるおだやかな味わいです。

This cultivar was selected from an Uji Zairai plantation (see p.40) and its prominent vegetal notes resembling green vegetables, gives it a distinct “green” impression. If Yabukita has a sharpness to it, the mouthfeel of Asatsuyu can be said to be more rounded and smooth.

[品種登録／年度] 農研機構果樹茶業研究部門 金谷／1953年
[Institution and registration year]
National Institute of Fruit Tree and Tea Science, Kanaya Japan / 1953

べにふうき | Benifuki

「べにふうき」はもともと紅茶用の品種として育種されました。紅茶らしい渋味とフルーティな香りを併せ持ちます。緑茶にした場合、渋味が強いですが、花粉症の症状を和らげると報告されているメチル化カテキンが残るため、緑茶でも飲まれています。

Benifuki was originally bred as a black tea cultivar and it certainly has typical black tea attributes such as the characteristic astringency and the peach like fruitiness. When made into green tea, the astringency becomes very prominent but on the other hand methylated catechin, which is reported to ease hay fever symptoms, goes unchanged which is one reason why it is consumed as a green tea as well.

[品種登録／年度] 農研機構果樹茶業研究部門 枕崎1993年
[Institution and registration year]
National Institute of Fruit Tree and Tea Science, Makurazaki Japan / 1993

Traveling Teapot —Tea and water quality
旅する急須 — 水のヒミツ

茶葉と茶器、そしてお湯か水さえあれば、いつでも、どこでも日本茶は楽しめます。毎日お茶が飲みたい私は海外出張や旅行の時でも必ず急須を持っていくことにしています。数年、地球を飛び回ってこの本を書いている現在、日本以外に15ヶ国で貴重なお茶の時間を過ごしてきました。実際にそれぞれの旅先でお茶を淹れてみると、同じ茶葉を使って同じ条件で淹れようとしても、やはり味と香りが変わってきます。ずっと同じ地域にいると気づかないかもしれませんが、淹れたお茶のおよそ99.5%が水で構成されていますので、必然的に香味はそれぞれの土地の水質によって変わります。お気に入りの茶器と茶葉とともに旅することによって、普段飲んでいるお茶を淹れても新しい発見に出会えます。お茶の香味は淹れ方も含めて様々なことに影響を受けますが、使う水に関しては軟水が日本茶を淹れるにはふさわしいとされます。日本の水はほとんどが軟水ですが、仮に硬水で日本茶を淹れた場合、うま味は出るのに爽やかな香味などが乏しく重たく感じることがあります。海外などで日本茶を淹れる場合はその場所の硬度を確かめて、ボトリングされた軟水のナチュラルウォーターを購入するか、軟水器などを利用するのがいいでしょう。

As long as you have tea leaves, tea ware and water (hot or cold) you can enjoy Japanese tea anywhere at any time. Craving several cups a day, I would never fail to bring my teapot with me regardless of whether I go somewhere on business or for leisure. During the last couple of years I have traveled to many countries, and at the time of printing this book, I have shared many precious tea moments with my small teapot from Tokoname in a total of sixteen countries including Japan. As expected, even if I use the same leaves and attempt to steep it under the same conditions, the taste and flavor change at the different destinations. Spending a long time in the same area makes it easy to forget, but water actually makes up 99.5% of the tea we drink, making this change inevitable. So by bringing your favorite tea and tea ware with you wherever you go, you are of course able to enjoy a relaxing teatime, but at the same time you can also make new discoveries with something that you drink on an everyday basis.

There are many factors, including the steeping method, that affect the taste of tea, but when it comes to water, soft water (water with a low level of hardness) is considered preferable. Almost all water in Japan is soft, but when steeped in an area with hard water the *umami* will dissolve well, yet the supposedly fresh aroma becomes weak, resulting in a tea that feels heavy rather than refreshing. If you are steeping Japanese tea in an area with hard water, it might be a good idea to use a water softener or to purchase bottled soft natural water.

一時硬水の変化 | Changes in temporary hard water

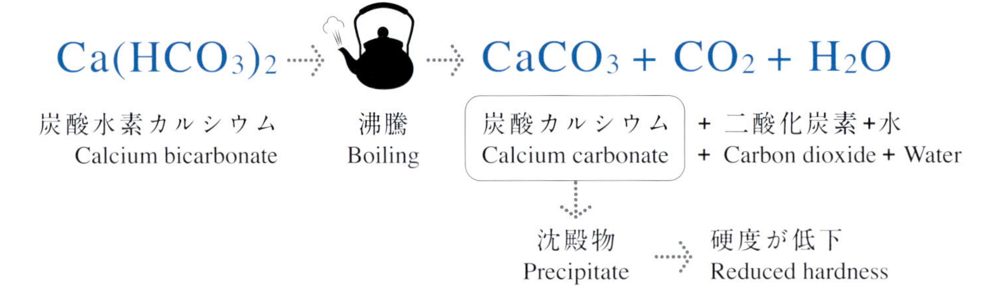

日本茶にふさわしい水があれば、正しい沸かし方もあります。緑茶は熱湯ではなく、70～80℃くらいで淹れた方が美味しいとされますが、沸騰させずにその温度にすれば良いと誤解している人もいるようです。沸騰したらそのまま3分程度おいて、お茶を淹れる際に必要な温度に下げるようにするのが基本です。あまり知られていませんが、沸騰させる理由が四つあります。一つ目は沸騰させることで塩素臭を抜くこと、二つ目は沸騰させて湯の中の酸素を追い出すこと、酸素を取り除くことでよりクリアな味わいになります。三つ目は水を殺菌し安全性を高めること、最後は沸騰させることによって硬度を下げることです。

硬水には「一時硬水」と「永久硬水」があります。前者は炭酸水素カルシウムや炭酸水素マグネシウムが硬度の主な成分で沸騰させるとこれらが沈殿物（炭酸カルシウムまたは炭酸マグネシウム）、二酸化炭素と水へと変わり、結果として硬度が下がります。とくに硬度が高い国で日本茶を淹れた時は怠ってしまうとせっかくの日本茶が台無しになることがあります。永久硬水は主に塩化カルシウムまたは塩化マグネシウムが硬度の理由のため、沸騰させても沈殿物が出来ず、硬度も下がりません。

In addition to there being a preferred water for Japanese tea, there is also an optimal way of boiling the water. Many types of green teas are in general considered to be steeped for the best result at 70-80°C (160°F-175°F). However, many seem to misunderstand this and think that you only need to heat up the water to the desired temperature without boiling it. In fact, the water should be boiled, left for about three minutes and then cooled down to the desired temperature before steeping. Although unknown even to many habitual tea drinkers, there are actually four reasons for boiling the water. First of all, it helps us to get rid of the smell of chlorine (if there is any). Second, it chases away oxygen from the water which gives the tea a clear taste. The third reason is to sterilize the water and make it safer. Finally, by boiling the water we can actually reduce its hardness.

Hardness can be divided into temporary hardness and permanent hardness. In the case of temporary hardness, mainly Calcium bicarbonate (also called Calcium hydrogencarbonate) and Magnesium bicarbonate (also called Magnesium hydrogencarbonate) are the molecules responsible. Boiling the water promotes the formation of precipitates (in the form of calcium carbonate and magnesium carbonate), carbon dioxide and water (H2O), and as a result we end up with softer water. Especially if you live in an area with hard or very hard water, it is important not to be negligent when it comes to boiling. Permanent hardness is usually caused by calcium chloride or magnesium chloride and unfortunately, as the name suggests, the hardness is permanent and cannot be reduced by boiling.

日本茶の産地
Japan’s tea growing regions

日本ではチャが、南の沖縄から北陸まで栽培されていますが、大きな産地は中部日本、関西、そして九州に集中しています。

それぞれの産地で異なる香味のお茶が作られており、煎茶の産地もあれば、玉露、かぶせ茶、そして中国式の釜炒り茶の産地もあります。静岡と京都は産地だけでなく、荒茶の仕上げと合組の技術を有する茶の集散地として際立っています。

Tea is grown, at least to some extent, from Okinawa in the south to the Hokuriku area, but most of the production is concentrated in Central Japan, the Kansai area, and the island of Kyushu.

All the different regions have their specific taste and flavor, and in many cases different tea types are produced, some specializing in *Sencha*, whereas others produce *Gyokuro*, *Kabusecha*, and some even Chinese style pan-fired tea. Two regions, Kyoto and Shizuoka, stand out for being not only tea producing areas, but also distribution centers where tea from many other regions is refined and blended.

栽培地域による茶葉の特性
Characteristics of different tea growing areas

平坦地の特徴
Characteristics of flat regions

- 深蒸し煎茶が中心 / Mainly *Fukamushi Sencha*
- 水色は濃緑色 / Deep green liquor
- 濃厚な地味 / Rich, round taste
- 飲みやすい / Smooth mouthfeel

山間地の特徴
Characteristics of mountainous regions

- 煎茶が中心 / Mainly *Sencha*
- 水色は山吹色 / Bright golden yellow liquor
- バランスが良い地味、香り高い / Balanced taste and strong aroma
- 切れの良い渋み / Slightly astringent

日本茶の歴史｜Timeline of Japanese Tea history

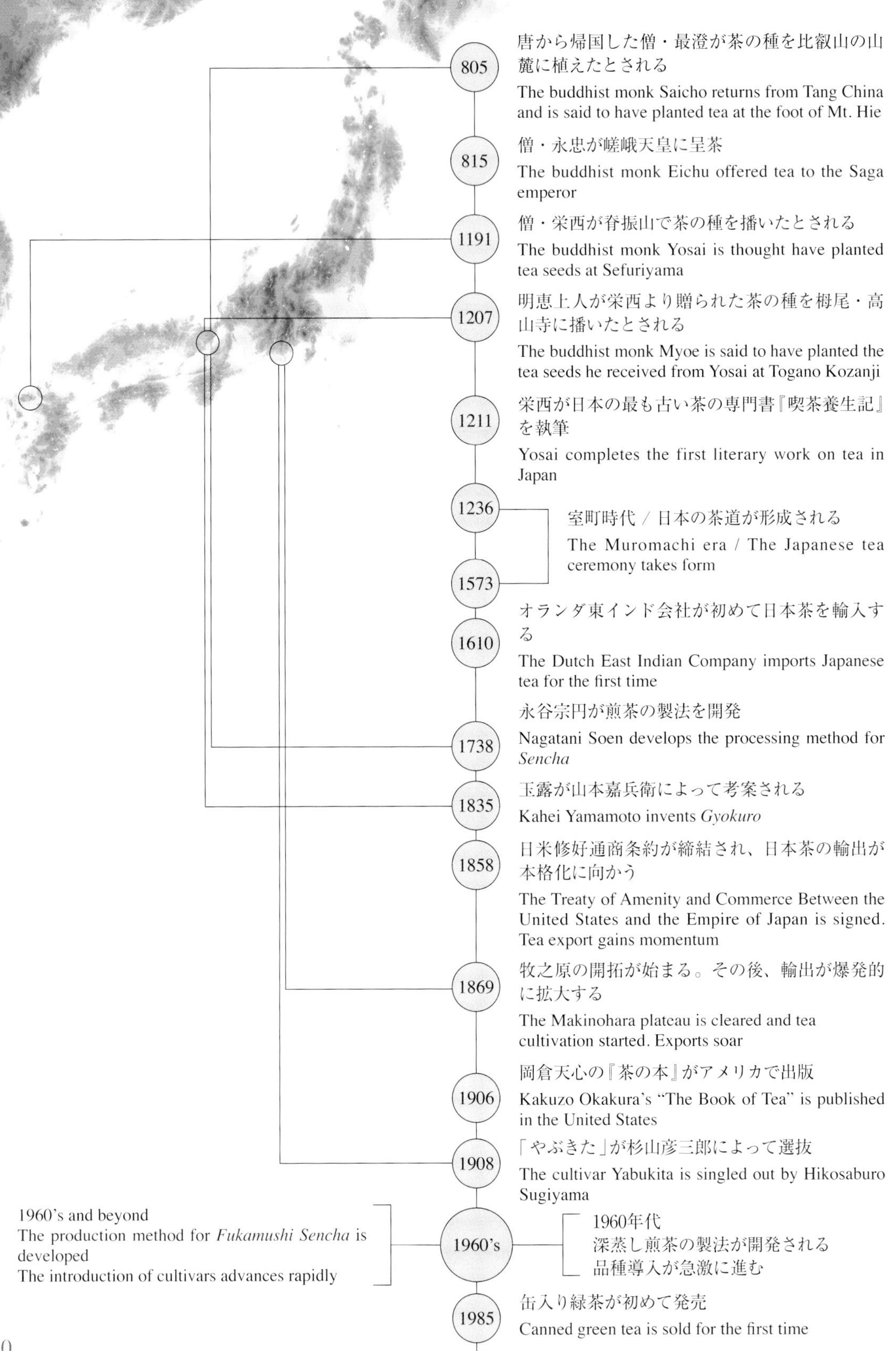

805
唐から帰国した僧・最澄が茶の種を比叡山の山麓に植えたとされる
The buddhist monk Saicho returns from Tang China and is said to have planted tea at the foot of Mt. Hie

815
僧・永忠が嵯峨天皇に呈茶
The buddhist monk Eichu offered tea to the Saga emperor

1191
僧・栄西が脊振山で茶の種を播いたとされる
The buddhist monk Yosai is thought have planted tea seeds at Sefuriyama

1207
明恵上人が栄西より贈られた茶の種を栂尾・高山寺に播いたとされる
The buddhist monk Myoe is said to have planted the tea seeds he received from Yosai at Togano Kozanji

1211
栄西が日本の最も古い茶の専門書『喫茶養生記』を執筆
Yosai completes the first literary work on tea in Japan

1236–1573
室町時代 / 日本の茶道が形成される
The Muromachi era / The Japanese tea ceremony takes form

1610
オランダ東インド会社が初めて日本茶を輸入する
The Dutch East Indian Company imports Japanese tea for the first time

1738
永谷宗円が煎茶の製法を開発
Nagatani Soen develops the processing method for *Sencha*

1835
玉露が山本嘉兵衛によって考案される
Kahei Yamamoto invents *Gyokuro*

1858
日米修好通商条約が締結され、日本茶の輸出が本格化に向かう
The Treaty of Amenity and Commerce Between the United States and the Empire of Japan is signed. Tea export gains momentum

1869
牧之原の開拓が始まる。その後、輸出が爆発的に拡大する
The Makinohara plateau is cleared and tea cultivation started. Exports soar

1906
岡倉天心の『茶の本』がアメリカで出版
Kakuzo Okakura's "The Book of Tea" is published in the United States

1908
「やぶきた」が杉山彦三郎によって選抜
The cultivar Yabukita is singled out by Hikosaburo Sugiyama

1960's
1960年代
深蒸し煎茶の製法が開発される
品種導入が急激に進む
1960's and beyond
The production method for *Fukamushi Sencha* is developed
The introduction of cultivars advances rapidly

1985
缶入り緑茶が初めて発売
Canned green tea is sold for the first time

煎茶の誕生 | The birth of Sencha

日本人にとって、現在の「煎茶」はあって当たり前の存在で、何百年も前からこの形で飲まれてきた伝統的な文化と思われがちです。しかし、実際には他の茶種よりも歴史が浅く、しかも大衆が日常的に飲むようになったのは戦後のことですから、つい最近であると言えます。

喫茶文化もチャも中国から渡来し、9世紀に唐から茶を煮出して飲む文化、12世紀に宋から抹茶法、17世紀になって明から急須を用いて茶を湯に浸してエキスを飲む喫茶法が、僧侶を中心に伝えられました。

千利休などによって茶道の文化が形成され、16世紀以降、抹茶は栽培、製造も含めて日本らしいものとなっていきましたが、煎茶道が成立するのは18世紀の江戸時代中期からで、現代まで茶器を含めて多分に中国趣味が残されてきました。

現在の煎茶の作り方のベースとなる製法が開発されるのは1738年のことで、完成したのは宇治田原の永谷宗円だと言われています。このお茶は生葉を蒸してから揉みながら乾燥させる製法であり、急須で淹れても味が出やすいお茶が、市場に出てくるのはこの頃になります。しかし、一般には煮出して飲む「煎じ茶」の方が主流でした。戦後になり、現在の「煎茶」の製造方法が確立し、一般大衆も享受する文化となりました。

こうして、ようやく本当の意味で日本らしいお茶が誕生しましたが、その進化が未だに進んでいる面白い時代に我々は生きています。

永谷宗円生家／Nagatani Soen's dwelling

In modern Japan, *Sencha* is something taken for granted, and many tend to believe that it has been consumed in the same way since many hundreds of years back. In fact, the history of *Sencha* is much shorter compared to many other types of tea, and it became a common beverage only after the second world war. Until then, most tea consumed in Japan was of different color, shape and taste.

Both the habit of drinking tea and the plant itself are believed to have been introduced from China, and Japanese monks played the main role bringing back different kinds of Chinese tea depending on the era in which they happened to live. From the Tang dynasty they imported the tradition of cooking and boiling of tea leaves, from the Song dynasty powdered tea, and finally the habit of using teapots to steep tea and drink the resulting infusion was imported from the Ming dynasty. Not only the tea itself, but also its cultivation and production took hold in Japan gradually.

Tea masters like Sen no Rikyu shaped the Japanese tea ceremony, thereby turning the powdered tea *Matcha* into something truly Japanese, both in terms of cultivation and processing. As for "*Senchado*", the way of *Sencha*, which was formed in the 18th century in the middle of the Edo period, the influences from China remains strong even today, something that can be seen in the tea ware used.

The processing method for *Sencha*, or at least the base for it, was developed in 1738, an achievement ascribed to Nagatani Soen from Ujitawara in modern day Kyoto prefecture. The new tea was steamed after plucking, and then dried while rolled and kneaded, and the taste and flavor dissolved easily even when steeped in a teapot. However, tea that was cooked and boiled to make an extract retained its position as a mainstream tea, and it was not until after the Second World War that *Sencha* secured its position as the typical tea consumed by the average Japanese.

This was how a truly Japanese leaf tea was finally born, and as its evolution continues on today, we might conclude that we do indeed live in exciting times.

日本茶の産地｜Japan's tea growing regions

静岡｜Shizuoka

拡大図｜enlarged map

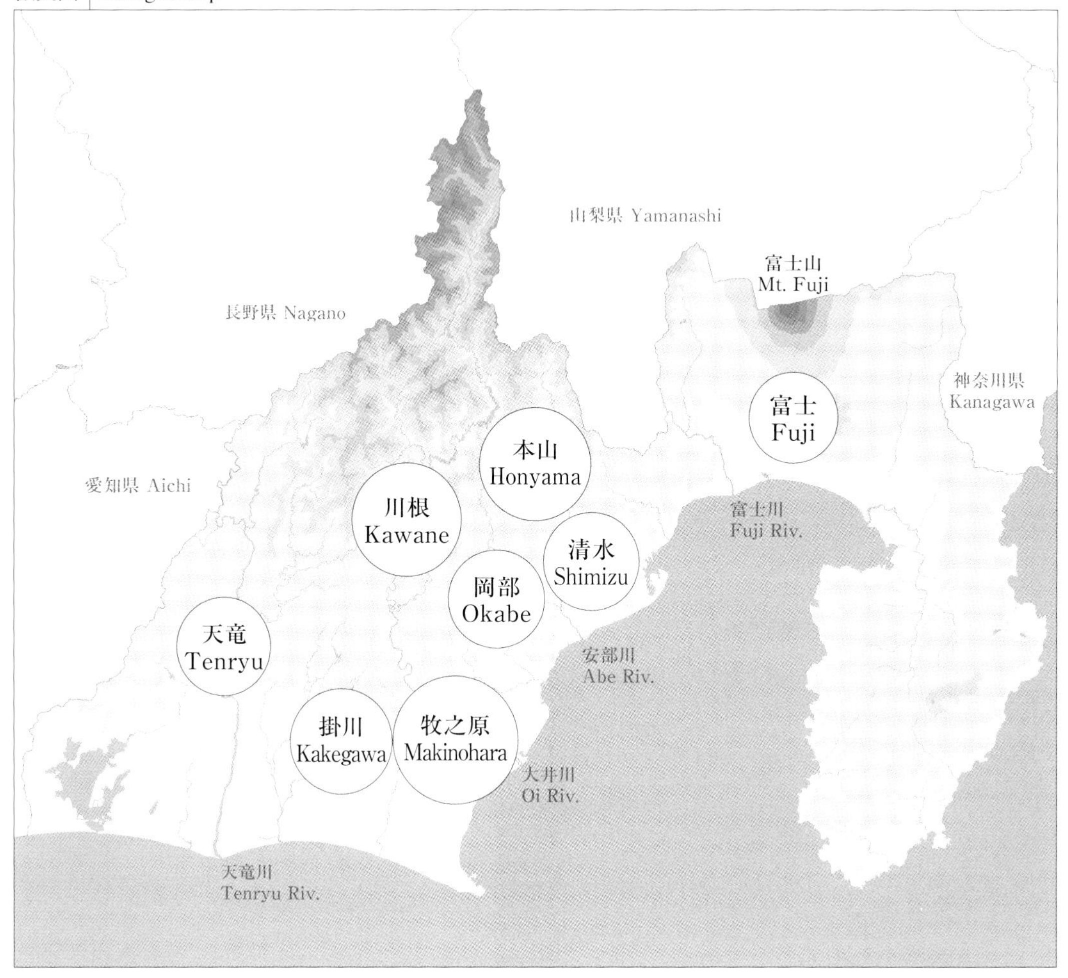

茶園面積でも生産量でも国内1位を誇る静岡は、美しい緑色の茶園の風景が新幹線からも見える一目瞭然の茶所です。「静岡茶」とひと言でまとめられますが、東西に長く、標高の差が大きな静岡はそれぞれの産地によってお茶の香味に違いがあります。平坦地で作られる「里のお茶」と山間地で作られる「山のお茶」とに大きく分けられますが、環境だけでなく、製法も異なっています。牧之原市や掛川市は深蒸し煎茶の産地として知られる一方、静岡市や島田市、浜松市の山間地では主に煎茶が作られるなど、同じ静岡でも大量生産から個性豊かなシングルオリジンの日本茶までと実に幅が広いです。しかも、産地としてだけではなく、京都と並び、仕上げと合組の技術を持つ茶問屋が多く集まるため、他産地の茶が仕入れられる集散地ともなっています。「やぶきた」をはじめ、全国で栽培されている多くの品種が静岡で育種されてきました。栽培面積は「やぶきた」が8割程を占めていますが、総生産量が多いため、「おくひかり」や「香駿」など複数の品種も多く栽培されています。

Shizuoka has both the largest cultivated area and production volume of all the tea growing regions of Japan. Even when seen from the bullet train, Shinkansen, it is easy to tell that this is a full-fledged tea region as the many tea plantations make up a beautiful green scenery. The tea often gets labelled as "Shizuoka tea" but Shizuoka stretches out far from east to west and with altitude ranging from high to low, all the teas from the respective subregions have their own distinct taste and flavor. It can be roughly divided into "*Sato no ocha*" (literally "tea from the village") which is grown in flat areas, and "*Yama no ocha*" ("mountain tea") which is, as the term suggests, grown in mountainous regions. Not only the environment, but also the processing methods differ. Comparably flat areas like Makinohara and Kakegawa mainly produce *Fukamushi Sencha*, whereas *Sencha* is the most common type of tea in mountainous regions like Shizuoka City, Shimada and Hamamatsu. Shizuoka shows a great variety as we find everything from mass production to single estate tea in the very same prefecture. Furthermore, Shizuoka is not only a tea producing region, but just like Kyoto, a place where tea is refined and blended. Tea refiners and wholesalers are gathered in large numbers, and as tea from other regions is traded and refined in large quantities, Shizuoka also functions as a large distribution center for Japanese tea. In addition, *Yabukita* and many other cultivars that are grown all over Japan have their origin in Shizuoka. In Shizuoka prefecture, Yabukita covers about 80% of all tea plantations, but since the total production volume is so great, considerable quantities of other cultivars, e.g. Okuhikari, Yamakai and Koshun etc, are also grown.

本山 | Honyama

安倍川の上流域の茶産地は本山と呼ばれ、典型的な「山のお茶」に当たります。水色は山吹色で、香味は爽やかな渋味に余韻が長く、「山のお茶」ならではの馥郁たる香りが特徴です。作業性が悪い急斜面の園地が多いため大量生産には向いていませんが、少量ながら高品質で独特なお茶として重宝されています。歴史も長く、鎌倉時代に中国・宋から茶の種を持ち帰った聖一国師が静岡で茶を広めたとされ、足久保に蒔いたのがその始まりとされています。

The tea growing region along the upper stream of the Abe river is commonly referred to as Honyama, and this tea is an excellent example of mountain tea. The color of the liquor is golden yellow and it has a pleasant astringency. The aftertaste lingers for a long time, and this, together with the sweet aroma that you can only get from mountain teas, is characteristic of Honyama tea. Since many of the tea plantations are located on steep slopes, the working environment is far from easy, which makes this area unfit for mass production. Instead, tea from this region is valued for its quality. The history is old, stretching back as far as the Kamakura period (1185-1333) when Shoichi Kokushi, a buddhist monk, brought tea seeds back from China and planted them in Shizuoka. This is said to have taken place in the village of Ashikubo, which is part of the Honyama area.

主な栽培品種：やぶきた、おくひかり、山峡
Main cultivars : Yabukita, Okuhikari, Yamakai

牧之原 | Makinohara

平坦地である牧之原台地を訪ねると見渡す限り茶園が広がっていますが、産地としての歴史は古くありません。明治維新によって失業した武士やインフラ整備の発展によって職を失われた川越人足が、荒地だった牧之原台地を開拓して茶栽培に取り組みました。それは、幕末から盛んになった茶の輸出の拡大を期待してのはじまりでした。現在、牧之原のお茶は国内消費が主体ですが、当時はほとんどが主な輸出先であったアメリカへ出荷されていました。牧之原では「やぶきた」を中心に深蒸し煎茶が作られてきましたが、品種の多様化の試みもあり、ここ数年「つゆひかり」という品種も増えてきました。

The first thing that strikes a visitor to the Makinohara plateau would definitely be the vast tea plantations stretching almost as far as the eye can see. However, despite the large cultivated area, Makinohara has a relatively short history as a tea growing region. During the Meiji restoration in the late 1800's, the samurai lost their privileges and the construction of bridges over Japan's many rivers made the workers who used to carry both things and people over the river redundant. These workers joined forces with the samurai in their great endeavor to clear up the wasteland that made up the Makinohara plateau, and thus found a new occupation as tea farmers. Tea export took off during the last years of the Tokugawa Shogunate, and the hope of increased revenues from trade with foreign countries was the main motivating factor for those who started up the tea estates in the area. Nowadays, most of the tea produced in Makinohara is consumed in Japan, but in the beginning almost all of it was shipped to the greatest importer of Japanese tea at the time, namely the United States of America. The main cultivar is Yabukita and the tea produced is essentially *Fukamushi Sencha*, but attempts to diversify the cultivation is being made, with the the cultivar Tsuyuhikari being grown to a larger degree in recent years.

主な栽培品種：やぶきた
Main cultivars : Yabukita

川根 | Kawane

本山に並び、川根も「山のお茶」の産地のひとつです。新芽がまだ柔らかい状態で摘まれ、品評会で見られる針のように細く縒れた形のお茶が多く見られます。品種は「やぶきた」が中心ですが、甘い香りと力強さが感じられる「おくひかり」なども栽培されています。

Tea from Kawane is, together with Honyama tea, usually referred to as mountain tea. The tea leaves are picked when they are still in a very soft and tender condition, and they tend to resemble the *Sencha* curled into fine needles often seen at official tea competitions. The most common cultivar is Yabukita but the slightly stronger Okuhikari with a notable sweet fragrance is also grown in this region.

主な栽培品種：やぶきた、おくひかり
Main cultivars : Yabukita, Okuhikari

掛川 | Kakegawa

生産量だけでなく、お茶の消費量も多い掛川は健康寿命が長いことで注目され、掛川茶の知名度が上がるきっかけにもなりました。「里のお茶」の産地に当たる掛川では主に「やぶきた」の深蒸し煎茶が作られています。他の産地よりやや香ばしいお茶が多いようです。

Tea is not only produced, but also consumed in large quantities in Kakegawa, an area that has received much attention for its residents long and healthy life expectancy. Since tea drinking is thought to be one of the possible contributing factors to longevity, this has made Kakegawa tea well known throughout Japan. Kakegawa tea belongs to the category village teas, and mainly *Fukamushi Sencha* is produced, not often with a slightly stronger aroma compared to other regions that produce similar teas.

主な栽培品種：やぶきた
Main cultivars : Yabukita

その他の産地 | Shizuoka - Other regions

静岡は、ほぼどこでもチャが栽培されています。前述の産地以外に「山のお茶」には清水と天竜などがあり、「里のお茶」としては菊川や袋井などが挙げられます。煎茶と深蒸し煎茶以外に玉露が生産されている岡部などもあります。そして日本を象徴する富士山の麓にも美しい茶園風景が広がっています。

Tea is grown throughout almost the whole of Shizuoka. Among mountain teas we find Shimizu and Tenryu, and among village teas we find Kikugawa, Fukuroi and many others. Most subregions produce either *Sencha* or *Fukamushi Sencha*, but Okabe stands out as a *Gyokuro* producing areas. Last but not least, one should not forget the picturesque tea plantations stretching out at the foot of Mt. Fuji.

日本茶の産地｜Japan's tea growing regions

京都・近畿圏｜Kyoto・Kinki region

拡大図｜enlarged map

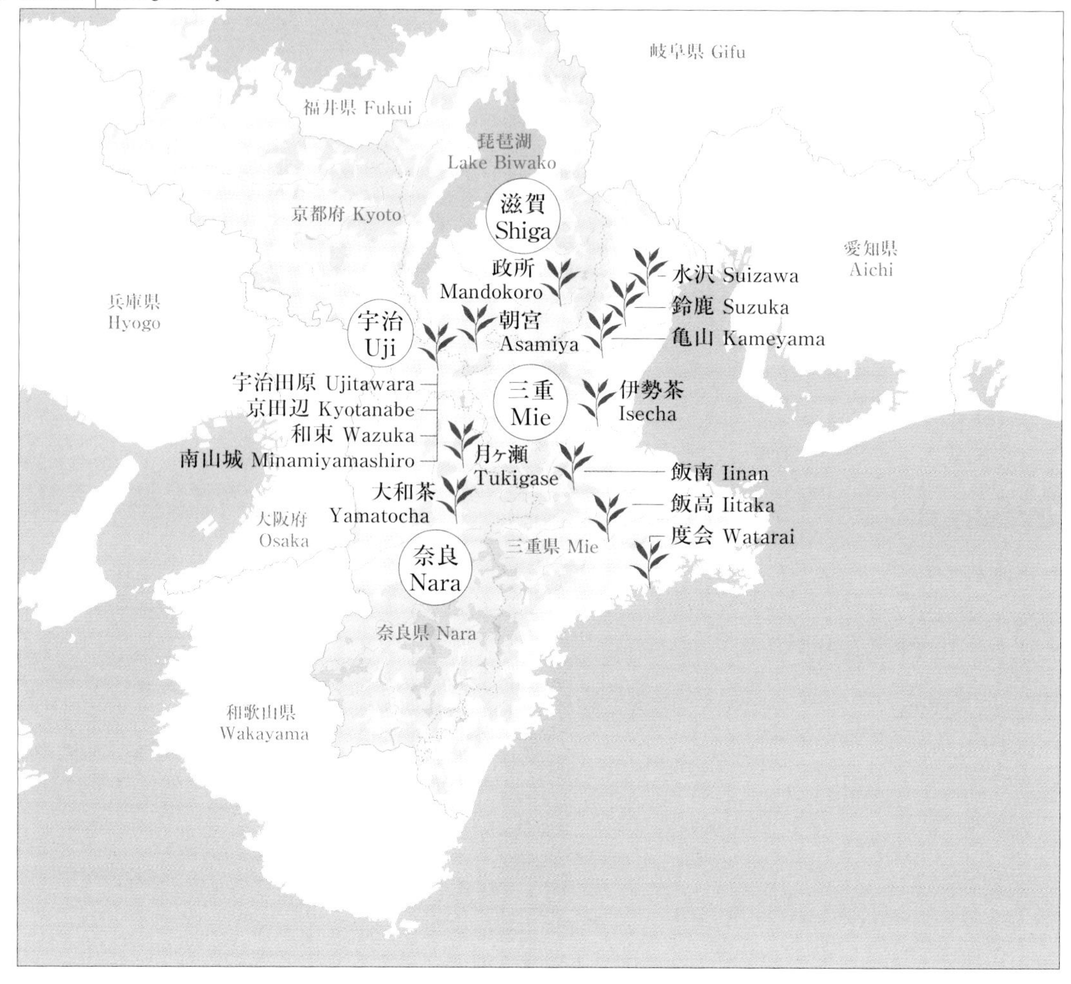

京都は抹茶の原料となる碾茶の生産量では1位を誇り、煎茶と玉露の発祥地でもあります。茶栽培だけではなく、他の産地の荒茶の仕上げと合組（ブレンド）を行う茶問屋が多く集まる集散地ともなっています。

静岡と鹿児島と比較すると生産量が少ないですが、多くの茶種が作られ、品種も多様化しています。特に被覆栽培の技術が発展し、碾茶や玉露などが最高級のお茶として評価されています。このような覆下栽培に向いている「あさひ」「さみどり」「ごこう」「うじひかり」などが京都で育種または選抜された品種です。東海地方や九州などには深蒸し煎茶の産地が多いですが、京都も含めて関西にはほとんどない茶種です。

茶産地としての歴史が長く、1191年に宋から帰国した栄西がチャの種子を京都の栂尾・高山寺の明恵上人に贈ったことがその始まりだとされています。産地を当てる闘茶会が流行っていた14世紀に栂尾のお茶は「本茶」とされ、それ以外のお茶は「非茶」とされたほど京都のお茶は昔から特別視されてきました。

Kyoto produces more *Tencha* (the raw material for *Matcha*) than any other prefecture, but it is also the origin of both *Sencha* and *Gyokuro*. Not only tea cultivation, but tea from other regions are refined and blended in Kyoto, thereby making it one of the main centers for tea distribution in Japan.
Compared to Shizuoka and Kagoshima the production volume is rather small, but many different kinds of tea are being produced and a wide variety of cultivars are grown. In particular the art of shading tea plants is well developed in Kyoto, and its *Tencha* and *Gyokuro* are praised as teas of the highest quality. Special cultivars for this purpose such as Asahi, Samidori, Goko and Ujihikari all have their origin in Kyoto. In the Tokai area and in Kyushu, *Fukamushi Sencha* is common, but in Kyoto and the rest of the Kansai area it is almost non-existant.
As a tea growing region, Kyoto has a long history stretching back to when Myoe, a monk at Togano Kozanji temple, received tea seeds from Yosai (also called Eisai), another monk who brought back tea seeds from Song dynasty China in 1191. In the 14th century, when tea tasting games were a popular pastime among the warrior class, Togano tea was seen as the real tea whereas all the rest were lumped together as "non-teas". Kyoto's unique position in the field of Japanese tea indeed goes back far in history.

宇治 | Uji

宇治茶はおそらく日本茶の最も知名度が高いブランドですが、実際には宇治だけでなく、京都府、滋賀県、奈良県、三重県産の茶を宇治の茶商がブレンドしたものも宇治茶とされます。宇治市内は碾茶と玉露の生産が中心となり、豊かなうま味と海苔を思わせるような「かぶせ香」が特徴です。昔は葦簀と藁で茶園をかぶせましたが、現在はほとんどの場合、寒冷紗と呼ばれる黒い色の化学繊維の布が使われます。

Uji is probably the most widely known tea brand from Japan, but in fact it consists of a blend of tea from four prefectures; Kyoto and the neighboring Shiga, Nara and Mie. Tea production in proper Uji is dominated by *Tencha* and *Gyokuro*, with an intense *umami* and a seaweed like aroma as the main characteristics. Traditionally the tea gardens were shaded with a reed screen and straw, but nowadays it is rare to see as most farmers use a black synthetic cloth instead.

主な栽培品種：さみどり、ごこう、あさひ
Main cultivars : Samidori, Goko, Asahi

その他の京都の産地 | Kyoto - Other regions

京都は宇治茶で有名ですが、府内で最も生産量が多いのは和束町です。他には玉露の生産が盛んな京田辺、そして南山城、城陽、八幡、宇治田原なども挙げられます。同じ京都府でも被覆栽培も露地栽培も多くの茶種が作られます。宇治茶としてではなく、ここ数年はそれぞれの産地のお茶として販売しようとする動きがあります。

Kyoto is famous for Uji tea, but the largest tea producing area in the prefecture is Wazuka. Among others we find Kyotanabe, famous for *Gyokuro*, and also Minami Yamashiro, Joyo, Yawata and Ujitawara. Although located in the same prefecture, many different teas are produced, both shaded and non-shaded. In recent years, more producers in these regions are attempting to sell their tea as single origin teas, rather than letting it be blended into *Ujicha*.

主な栽培品種：やぶきた、おくみどり、在来
Main cultivars : Yabukita, Okumidori, Zairai

滋賀 | Shiga

805年に唐から帰国した最澄が比叡山の麓にチャの種子を植えた伝説がありますが、それが現在、滋賀県大津市にある日吉茶園だと伝えられています。滋賀県のお茶は多くの場合、宇治でブレンドされますが、朝宮や政所という産地はブレンドされずに流通することもあります。在来の茶園（40頁参照）が多く残っていることから特に政所茶は希少な存在として重宝されています。

According to legend, Saicho, a monk returning from Tang dynasty China in 805, planted tea seeds at the foot of Mt. Hie, and the small Hiyoshi tea garden that can be found in Otsu City in Shiga prefecture may be the remains of this site. Most of the tea produced in Shiga is transported to Uji for blending, but some of the tea from subregions like Asamiya and Mandokoro are also sold as single origin teas. Especially Mandokoro, where many old *zairai* plantations (see p.40) can be found, is highly valued as a rare tea region.

主な栽培品種：やぶきた、在来
Main cultivars : Yabukita, Zairai

奈良 | Nara

県内の大きな産地である月ヶ瀬は京都に隣接していることもあり、京都でも関西全体でも好まれる甘味が強く、海苔のような香りがするかぶせ茶の生産が多い産地です。奈良の多くのお茶は宇治でブレンドされますが、大和茶というブランド名で流通することもあり、月ヶ瀬などシングルオリジンのお茶として売る動きもあります。

Nara produces mainly *Kabusecha*, a type of tea that with its sweet taste and distinct seaweed like aroma, is popular both in Kyoto and the Kansai region as a whole. As with other teas grown in the Kansai region, most Nara tea also gets blended in Uji, but some of it is sold under the brand name *Yamatocha*, or even as single origin tea from subregions like Tsukigase.

主な栽培品種：やぶきた、おくみどり、在来
Main cultivars : Yabukita, Okumidori, Zairai

三重 | Mie

静岡県と鹿児島県に次いで生産量で3位を誇る三重県ですが、「三重のお茶」としてよりも、その多くが京都の宇治でブレンドされて「宇治茶」として販売されています。主にかぶせ茶が作られているため、新茶が摘採される時期に茶園を訪ねれば、ほぼ真っ黒の寒冷紗で茶園が覆われています。三重県産のお茶の一部は「伊勢茶」というブランド名でも流通しています。三重県はお茶の産地だけでなく、四日市で作られている万古焼の急須も有名です。

Mie is Japan's third largest tea producer after Shizuoka and Kagoshima, but since most of the tea gets blended in Kyoto as Ujicha, the likelihood of coming across pure Mie Tea is rather low. *Kabusecha* is the most common type, and if one were to visit a tea farm during the harvest season, the lush green hedges would most likely be completely covered by the synthetic black cloth most commonly used. Some of the tea produced in Mie is sold as Isecha. Mie is not only a tea growing region, but is also famous for Yokkaichi Banko teapots.

主な栽培品種：やぶきた、在来
Main cultivars : Yabukita, Zairai

日本茶の産地｜Japan's tea growing regions

東日本・東海｜East Japan Tokai area

拡大図｜enlarged map

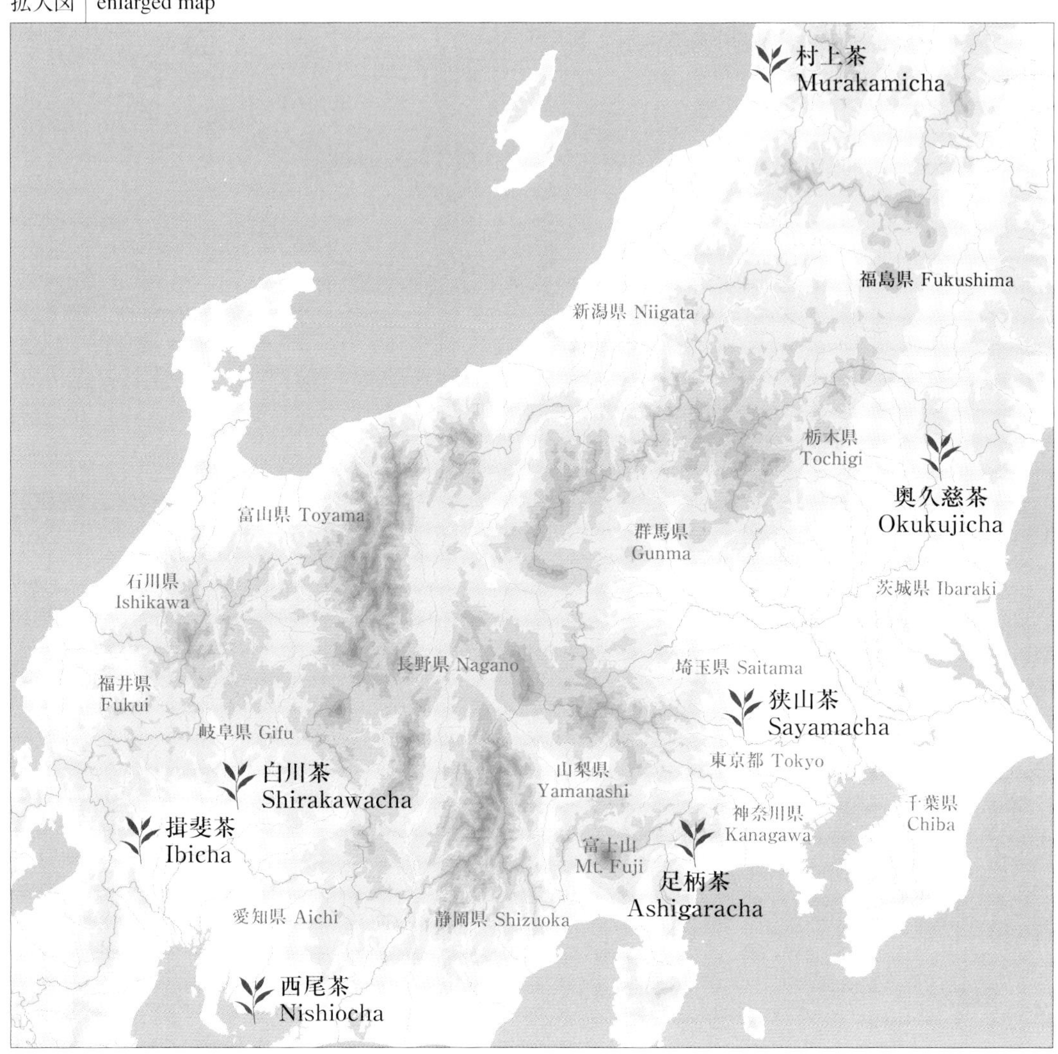

狭山｜Sayama

狭山茶は埼玉県で作られているお茶の総称ですが、長い歴史を持つ産地でもあります。起源が不確かですが、南北時代に書かれた資料に武蔵川越が表記されていることから少なくとも700年の歴史があることがわかります。現在では主に「やぶきた」の深蒸し煎茶が作られています。

The brand name Sayama tea includes all tea made in Saitama prefecture and, although small, it is one of the oldest tea growing regions in Japan. Its origin is unclear but the earliest reliable records where Musashi Kawagoe in Saitama is mentioned as a tea growing region date back to the Nanboku era (1336-92). *Today Fukamushi Sencha* made from Yabukita is the most common type of tea.

主な栽培品種：やぶきた、さやまかおり、ふくみどり
Main cultivars : Yabukita, Sayamakaori, Fukumidori

西尾 | Nishio

愛知県の西尾では碾茶の生産が盛んで、京都に次いで生産量2位を誇ります。愛知県も静岡と同様に僧侶の聖一国師が鎌倉時代にもたらしたチャの種子が起源とされています。ただし、茶の生産が本格的に拡大されたのは明治時代に入ってからです。その際には京都の製茶技術が導入されました。

Nishio in Aichi prefecture almost exclusively produces *Tencha*, and after Kyoto it boasts the second largest production volume. Just like neighboring Shizuoka, the monk Shoichi Kokushi is usually credited for introducing tea to Aichi prefecture in the Kamakura era. However, it would take until the Meiji era, when tea making skills were introduced from Kyoto, for tea production to really take off in Nishio.

主な栽培品種：やぶきた、さみどり
Main cultivars : Yabukita, Samidori

岐阜 | Gifu

岐阜県には白川茶や揖斐茶などがありますが、毎年冬に雪が積もることから他の産地と比べて比較的寒いことがわかります。揖斐川沿いにある春日は在来の茶園(40頁参照)が多く残っている珍しい産地です。白川は主に「やぶきた」が栽培され、形状が細かく、水色が濃い緑茶がほとんどです。

Ibi and Shirakawa are two subregions of Gifu, and their snow covered tea plantations in winter reveal how it is colder than most other tea regions in Japan. Kasuga along the Ibi river is a rare region as many old *zairai* plantations (see p.40) still remain. In the Shirakawa area, Yabukita is the most commonly grown cultivar, and the tea has a slightly broken appearance that when steeped gives the liquor a thick green color.

主な栽培品種：
やぶきた、おくみどり、在来
Main cultivars :
Yabukita, Okumidori, Zairai

東日本・東海地方のその他の産地 | East Japan, Tokai area and other regions

大規模ではありませんが、他の県でもチャの栽培が行われています。日本で経済的な茶栽培の北限とされているのは新潟県の村上市と茨城県の大子町を繋いだ線です。他にも、神奈川県、千葉県などに茶産地があります。

There are many regions in Japan that produce tea on a small scale, but the line between Murakami in Nigata prefecture and Daigo in Ibaraki prefecture is considered to be the northern limit for commercial tea farming in Japan. Although only small volumes, tea is also grown in Kanagawa prefecture and Chiba prefecture.

日本茶の産地 | Japan's tea growing regions

九州・四国 | Kyushu Shikoku

拡大図 enlarged map

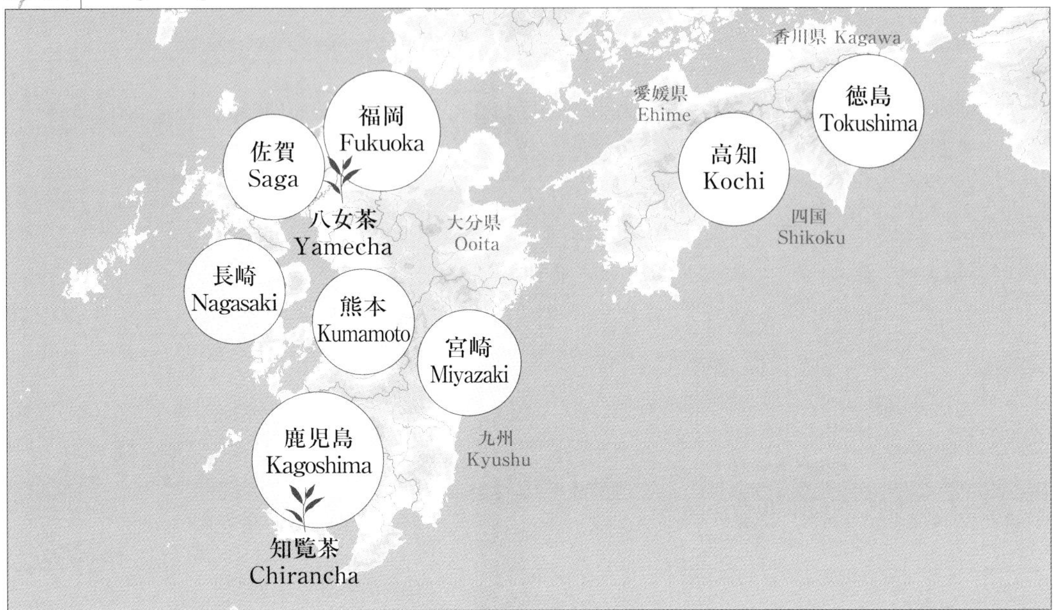

四国の産地 | Shikoku

四国の生産量は少ないですが、日本では極めて珍しい微生物の働きを活かした後発酵茶の生産で有名です。この中には高知県の碁石茶（31頁参照）、徳島県の阿波番茶、愛媛県の石鎚山の黒茶などがありますが、どれも一般的な日本茶と大きく異なり、強い酸味と独特な土を思わせる香りが特徴となります。

The island of Shikoku only produce small quantities of tea compared to other regions, but is famous as the home of teas fermented by micro-organisms, something very rare in Japan. We find examples of teas like this in *Goishicha* (see p.31) from Kochi, *Awabancha* from Tokushima and *Ishizuchi san kurocha* from Ehime. All of these are very different from Japanese tea in general, having a strong sourness and a peculiar earthy taste.

福岡 | Fukuoka

鎌倉時代に福岡県と佐賀県との境にある背振山に栄西が茶園を開いたとされていますが、このように歴史が長い福岡は現在、八女茶という深蒸し煎茶のブランドのほかに玉露の産地としても有名です。一部の玉露用品種を除けば、「やぶきた」が中心ですが、ここ数年は「つゆひかり」なども少しずつ普及してきました。

The oldest tea plantation in Fukuoka dates back to the Kamakura era, when Yosai (also called Eisai) is said to have cultivated tea at Sefuriyama on the border of Saga prefecture. Fukuoka not only has a long history, but has earned fame for Yame tea, a *Fukamushi Sencha* brand, and for the production of *Gyokuro*. Apart from the cultivars grown for making *Gyokuro*, Yabukita dominates, but in recent years other cultivars like Tsuyuhikari are also gaining ground.

主な栽培品種：やぶきた、かなやみどり、さえみどり
Main cultivars : Yabukita, Kanayamidori, Saemidori

鹿児島 | Kagoshima

全国2位の生産量を誇る鹿児島は、かつて大規模な紅茶の産地でした。インドやスリランカの大量生産により国際市場での競争力が低下し、また、1970年代の紅茶の輸入自由化のため国内でも外国産の紅茶が主流となりましたが、同時期の緑茶の内需の高まりにより、紅茶から緑茶へと生産を転換しました。平坦地に見渡す限り広がる大規模な茶園は鹿児島ならではの風景です。気候が他の産地より温暖のため新茶が早く、４番茶と秋冬番茶まで豊富に摘み採れます。日照量の多い産地のお茶は渋味成分の含有量が高くなりがちですが、甘味を保つために摘採の前に茶園に覆いをかぶせることがあり、品種も摘みとり時期の早い「ゆたかみどり」「さえみどり」「あさつゆ」などが普及しています。ここ数年までは荒茶を静岡などに出荷していましたが、最近では知覧茶などのブランドの知名度が高まり、純粋な鹿児島茶としても販売されるようになりました。また、ここ数年は輸出に力を入れているため有機栽培茶が増えてきました。

Kagoshima holds the second position in terms of both cultivated area as well as production volumes, but few know that it used to be a large scale black tea region. However, due to mass production of black tea in countries like India and Sri Lanka, Japanese black tea could not compete on the international market. The decisive blow came with trade liberalization in the 1970's, when import tariffs on black tea in Japan disappeared, thereby pushing Japanese tea out of the domestic market in favor of tea from other countries. However, during the same period, domestic demand for green tea rose, and Kagoshima managed to switch to green tea production. The vast tea plantations stretching out as far as the eye can see over large flat areas is a scene unique to Kagoshima. The climate is warmer than other tea growing regions in Japan so Shincha is harvested early, and even Yonbancha (the fourth harvest) and Shutoban (fall and winter harvest) tea are harvested in considerable quantities. Tea grown in regions with strong sunlight like Kagoshima tend to become bitter and astringent, so in order to preserve the natural sweetness of the tea, the tea plants are often shaded and cultivars low in astringency, e.g. Yutakamidori, Saemidori and Asatsuyu are cultivated to a larger degree than in other regions. As a result Kagoshima tea is rather smooth and sweet in taste. Until only a couple of years ago, Kagoshima tea was mainly shipped to Shizuoka for refining but in recent years pure Kagoshima tea is getting sold in larger quantities and brands like Chiran Tea are becoming increasingly known to the Japanese public. Kagoshima is also making great efforts to increase exports, and because of this organic farming is growing rapidly.

主な栽培品種：やぶきた、ゆたかみどり、さえみどり、あさつゆ
Main cultivars : Yabukita, Yutakamidori, Saemidori, Asatsuyu

その他の九州の産地 | Kyushu - Other regions

茶は九州全体で栽培されていますが、宮崎県の高千穂や佐賀県の嬉野などが中国式の釜炒り茶で有名な産地です。蒸し製緑茶と違い、あっさりした喉越しに甘い釜香が特徴です。

Tea is produced on the entire island of Kyushu, but some regions like Takachiho in Miyazaki prefecture and Ureshino in Saga prefecture are famous for Chinese style pan-fired teas. Their refreshing smoothness and sweet roasted aroma make them very different from steamed Japanese tea.

First, second and third flush Sencha
一番茶・二番茶・三番茶について

煎茶／ *Sencha*　　番茶／ *Bancha*

紅茶や烏龍茶などは一般的に時期によってそれぞれに味の特徴があると考えられていますが、日本の蒸し製緑茶は「一番茶」が飛び抜けて高品質なものとなっています。これには幾つかの理由があります。冬を越え栄養を蓄えて春に成長する新芽には、うま味成分が豊富に含まれています。また、新芽が均一に伸びやすく原葉を揃えやすいので、内容も姿も良いお茶に加工する可能性が高まります。

「二番茶」以降はうま味の成分も減り、葉の硬化や、伸び方も「一番茶」ほど均一にならないため、出来上がったお茶は艶がない裏側が表面に出たり、やや扁平（へんぺい）な形の茶葉になったりします。結果として「一番茶」ほど濃厚な味わいが出ません。「三番茶」や「四番茶」以降は葉がさらに固くなりますので、「三番茶」以降はペットボトル茶などの原料やほうじ茶ほか、安価なお茶となります。畑の管理と製茶技術によっては「二番茶」でも良いお茶を作ることは可能です。うま味が少ない分、さっぱりとした味わいでやや熱めのお湯で淹れても美味しいお茶にになります。

When it comes to black tea or Oolong tea, tea from different seasons are appreciated for their respective characters. However, in the case of steamed Japanese tea, *Ichibancha* (first harvest tea) is considered to be outstanding and the quality incomparable to the rest of the harvests. There are several reasons for this. First of all, after waking up from winter dormancy, all the absorbed nutrients will result in soft, tender buds rich in *umami* compounds. Also, since the spring buds grow at an even pace it becomes easier to pick leaves with a similar shape and size. This makes it possible to roll them into the characteristic needle like shape, thereby turning the leaves into a high quality *Sencha* that both looks beautiful and tastes excellent.

Nibancha (second harvest tea) and later harvests, are weaker in *umami*, the leaves also become more fibrous and hard, and do not grow as evenly as *Ichibancha*. After processing, the non-glossy side end up as the surface of the dried leaves, that also tend to get a slightly flat appearance. As a consequence, we are not able to get the same rich taste as we do from *Ichibancha*. *Sanbancha* (third harvest tea) and *Yonbancha* (fourth harvest tea) get even more fibrous and hard and is traded for a lower price. Tea from these harvests are in general used for making *Hojicha*, or as a base for making the tea beverages sold in plastic bottles that are widely consumed in Japan. However, if the plantation has been taken care of well, and the tea producer is skilled, even *Nibancha* can amount to a good taste experience. Since it has less *umami*, it can be enjoyed as a light and refreshing tea, tasting good even when steeped in slightly hot water.

茶葉の種類別 美味しい淹れ方
How to steep Japanese tea

日本の蒸し製緑茶は淹れ幅が最も広く自由度の高いお茶ですが、それゆえに初心者には多少難しいお茶かも知れません。成分が他の茶種よりも溶出しやすいので、苦味や渋味が出過ぎないようにお湯の温度と浸出時間に注意する必要があります。お茶を淹れる時に香味に影響を与えるのは主に茶葉の量と湯量、温度、そして浸出時間の4点です。実際には正しい淹れ方が存在しないので、マニュアル通りに淹れるよりも淹れ方の理論を最大限に生かして実験や料理を作るような気持ちで日本茶ならではのバリエーションを引き出して味わいましょう。

No other type of tea can be steeped in as many ways as steamed Japanese tea, but since it easily turns bitter, it can also be difficult for the beginner. Admittedly, its components are released a lot faster than other types of tea, so it is important to pay attention to both the temperature and the steeping time. When steeping tea, there are four main factors that affect the taste; amount of tea leaves, amount of water, temperature and steeping time. There is no correct way of steeping tea, so rather than just following the manual it is much more enjoyable to remember and then apply the logic of tea steeping to bring out the most of the great variation in taste that Japanese tea has to offer.

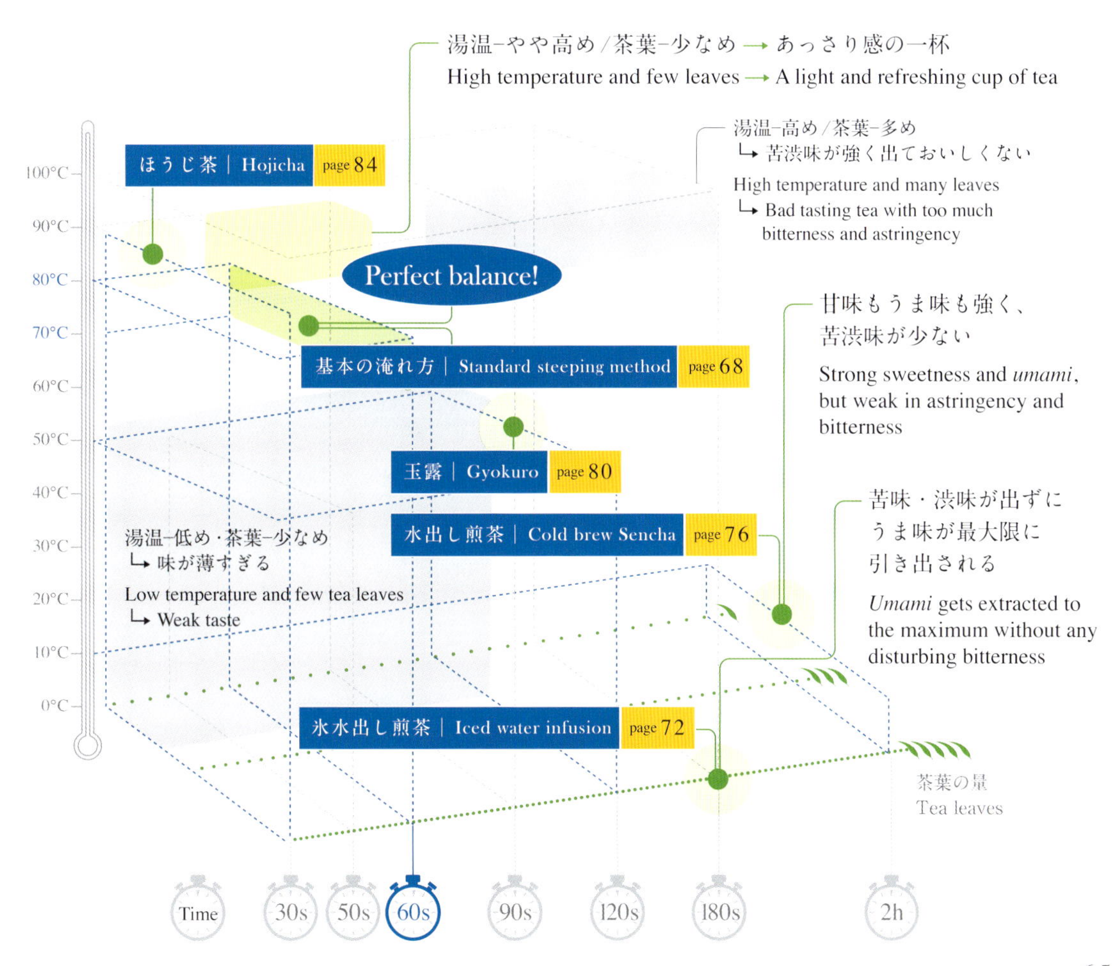

基本の道具 | Basic tea ware

1 茶さじ | Tea scoop

日常的にお茶を淹れる人なら必ず持っておきたいものです。茶葉量をおおよそ量る計量器ともなります。

For measuring. A necessity for anyone steeping tea on a daily basis.

2 茶筒・茶缶 | Tea Canister

なるべく密封性が高いものが良いですが、長期保存には向いていないので、茶筒に入れた茶は早めに消費しましょう。高温にならない場所の保管で約1ヶ月が目安です。

The tea caddie should be as airtight and possible, but is still not good for storing tea over a long period, so make sure to consume it as quickly as possible. If stored in a cool dry place, Japanese tea can be kept in a tea caddie for about one month.

3 湯呑み | Teacup

湯温を下げて淹れる煎茶や玉露には磁器、熱めでたっぷりと飲むほうじ茶や番茶には厚手の陶器などを選ぶといいでしょう。

White porcelain is good for teas steeped after cooling down the water, such as Sencha and Gyokuro. For teas drunk hot and in large quantities like Hojicha or Bancha, thick ceramic cups would make a better choice.

4 急須 | Teapot

詳しくは87頁「急須の楽しみ」へ
The beauty of Japanese Teapots → page 87

5 湯冷まし | Water cooler

一度沸かしたお湯を冷ますために用いる大事な道具です。

Used for cooling down the boiled water, the "*yuzamashi*" is an important vessel for making Japanese tea.

6 蓋置 | Lid rest

必須ではありませんが、あるなら安心して蓋を置けます。

Not necessary, but makes it easier to keep the lid safe when not in use.

茶葉の掬い方 | How to scoop tea

なるべく茶葉を均一に掬いたいため、ただ上部から取るのではなく、掬い方にも気を配るとバランスの良いお茶を淹れることが出来ます。まず、茶筒を少し斜めにし、逆さにした茶さじを茶筒の口の上側に差し入れます。その後、茶筒を内向き、茶さじを外向きに回します。そうすることで壊れることなく自然に茶葉が茶さじに落ちてきます。

As the aim is to get an average sample of the tea leaves, it is important to not just take from the top. Paying attention to how the tea is scooped will result in a well balanced cup of tea. First, tilt the tea caddie a bit and then hold the tea measure upside down and insert it in the upper part of the caddie. Then twist the caddie inwards and the tea spoon outwards. As a result of this motion, tea leaves will fall evenly on the spoon without getting crushed or broken.

急須のお手入れ | Teapot maintenance

急須はお茶を淹れ終えたら速やかに茶殻を捨て、水洗いし、自然乾燥させます。茶漉しが詰まったり、汚れが目立った場合は簡単に手入れが出来ます。急須専用のブラシを用意して茶漉しを綺麗にします。汚れはメラミンスポンジを使ったり、ひどい場合は重曹などを利用しましょう。吸水性の無い素材であれば酸素系漂白剤も使用出来ます。深蒸し煎茶や粉茶のような細かいお茶の場合、茶漉しが詰まりやすく、手入れも煩雑ですので、広い金網や茶籠付きの急須を選びましょう。

When you are done steeping your tea, throw away the used leaves, rinse the teapot and let it air-dry. Even if the strainer gets clogged or the stains stand out too much, there are easy ways to take care of a Japanese teapot. A brush can be used to clean the strainer. Light stains can be cleaned with a melamine foam sponge, and tough ones by using baking soda. For teapots made of non-porous materials oxygen-based bleach can also be used. Teas with a broken appearance like *Fukamushi Sencha* and *Konacha* tend to clog the strainer so maintenance can be troublesome. For these teas, teapots with a wide stainless strainer, or a strainer basket are preferable.

日本茶の保存方法 | How to store tea Japanese tea

お茶は光、酸素、湿気、熱、そして移り香で劣化し、変質してしまいます。開封後は密閉性の高い茶筒に移し冷暗所に置きましょう。未開封の茶袋は冷蔵庫でも構いませんが、開封時に結露しないように、開ける前には必ず常温に戻します。同じ理由で一度開けた茶袋を冷蔵庫に出し入れしないようにしましょう。

When exposed to light, oxygen, moisture, heat or strong smell the quality of tea easily degrades or alters. After opening your tea package, make sure to store the leaves in an airtight container in a dark, cool and dry place. If unopened, tea to can be stored in the fridge, but make sure to let it return to room temperature before opening it to protect the leaves from condensation. For the same reason, once the package is opened, tea should never be stored in the fridge.

|煎茶|基本の淹れ方
The standard steeping method for Sencha

煎茶の味の要素をバランスよく引き出した淹れ方です
How to steep a well balanced cup of Sencha

1

まず6gの茶葉を急須を入れます。茶さじならだいたい2杯、小ぶりのティースプーンなら3杯が目安です。まずこの分量で淹れてみて、好みに応じて茶葉を増減しましょう。

First, put 6g of *Sencha* leaves in the teapot, the equivalent of two full Japanese tea measures or about three teaspoons. After trying this once, add or decrease the amount of leaves to your own taste next time.

2

苦味や渋味、雑味などが出ないようにお湯を冷まします。ケトルから湯冷ましに注ぐと、温度は10℃ほど下がります。湯冷ましがなければ耐熱性のあるマグカップなどでも代用が可能です。ただ、湯冷ましを使った方が機能性も雰囲気も味わえますので、持っておくと良いでしょう。

Cool down the water to avoid excess bitterness, astringency and harsh tastes. By pouring boiling water from the kettle into the *yuzamashi*, the temperature drops with about 10°C. If you do not have a *yuzamashi*, any heat resistant vessel can be used. However, a *yuzamashi* is both functional and adds to the feeling so it is definitely a good idea to have one.

その他の適した茶葉

深蒸し煎茶（19頁）※70度 − 40秒
かぶせ茶（21頁）/ 茎茶、棒茶（25頁）
釜炒り茶（27頁）※80度

Other suitable teas

Fukamushi Sencha (p.19)
*70°C (160°F) − 40 seconds
Kabusecha (p.21) / Kukicha, Bocha (p.25)
Kamairicha (p.27) *80°C (175°F)

淹れ方	Steeping guide
茶葉の量　6g	Tea leaves　6g
湯量　180ml	Water　180ml
湯温　70 〜 80度	Temperature　70 〜 80°C (160 〜 175°F)
時間　60秒	Steeping time 60 seconds

3

70 〜 80℃くらいの湯温が煎茶の香味を最もバランスよく引き出します。湯を注いで少し待ってから器の底に手を当て、「触れるけれど熱く感じて長くは持てない」くらいが約70℃です。何回か温度計を使用して体感する温度を確認すれば感覚でわかるようになります。

In order to end up with a good balance of all the taste elements, the water should be cooled down to 70°C-80°C (160-175°F). Pour hot water into the *yuzamashi*, wait for a while and then try to hold it carefully. At 70°C (160°F), most of us can bare touching it for a while, but it is still too hot to hold for a long time. Using a thermometer a couple of times to ascertain the temperature will make you able to trust your senses from thereafter.

4

お湯の温度が十分下がったら静かに急須に注ぎます。温度が低すぎると味が薄くなったり香りが弱くなったりするので、冷めすぎないように蓋をして、約1分間待ちます。時間が短かければ味が出きっていない薄いお茶になり、逆に長ければ渋味などが出ます。

When the water has cooled down to the desired temperature, pour it gently into the teapot. If the temperature is too low, the tea will get a watery taste and the aroma will also become weaker. To prevent this, make sure to put on the lid, and then wait for one minute. If the steeping time is too short, the taste will not be fully extracted, but steeping the tea for too long will make it bitter and astringent.

｜煎茶｜基本の淹れ方

The standard steeping method for Sencha

5

激しく急須を振ったりすると雑味が出やすくなります。静かに注ぎましょう。２人分以上を淹れる時は濃さを調整するようにそれぞれの湯呑みに少しずつ廻し注ぎます。二煎目も美味しく味わえるように最後の一滴まで注ぎ切ります。

Pouring energetically will only result in a harsh taste, so handle the teapot gently and pour quietly to get a cup of tea with a smooth and round taste. If you are steeping tea for more than one person, pour little by little into each cup to adjust the strength and color. Always make sure to pour to the last drop.

6

お茶を注ぎ切り、お湯が残っていなくても急須の中は熱くなっています。茶葉が蒸れずに二煎目も美味しく味わえるように必ず蓋をとりましょう。蓋は蓋置きに乗せるか、あるいは注ぎ口の下に挟んだりして安全な場所に置きます。

Even if all the tea is poured and no liquid remains, the teapot is still hot. This lingering heat will make the second steeping excessively bitter and much of the aroma will also be lost, so make sure to remove the lid and put it on a lid rest or somewhere else safe.

二煎目を味わう

お茶を淹れる時に一番早く出るのがうま味と甘味ですが、二煎目以降はあっさり感があって爽やかな渋味と上品な苦味にお茶の香りがより強く立ち上がってきます。これらの要素を最大限に楽しむために徐々にお湯の温度を高めていきます。ただ、お茶の葉はすでに開いていますので、浸出時間を数秒だけにして待たずに湯呑みに注ぎます。お菓子をお茶と合わせるならキレのある渋味を持つ二煎目から味わうのが一番良いでしょう。

How to enjoy the second steeping

Most of the *umami* comes out in the first steeping, and in order to extract more of the aroma and a pleasant astringency from the second steeping, gradually raise the temperature. However, the leaves have already opened up so steep the tea for only a few seconds before pouring it into the teacups. If you are pairing your tea with sweets, the naturally astringent second and following steepings will make a perfect match.

味を均等にするために

2人または3人分を淹れる時、香味を均一にするために廻し注ぐのが大事なポイントです。ただ、淹れる時間も長くなってしまうので、4人以上の場合は温めてあるサーバーを使うと便利です。道具を増やさないように湯冷ましをサーバーとして使うのが良いでしょう。

How to adjust the taste evenly

When steeping tea for two or three people, it is important to pour little by little to make all the cups of tea taste the same. However, this action itself will make the steeping time longer, so pouring everything into a preheated pitcher or back into the Yuzamashi makes it easier to control the taste and flavor when steeping tea for a large number of people. Then pour the tea from the pitcher into the teacups.

|煎茶| 氷水を使った淹れ方
Steeping Sencha in iced water

うま味を最大限に引き出した日本茶の最も贅沢な味わい方の一つです
How to extract and indulge in a burst of umami

1

まずは氷水を用意します。氷も水も塩素臭がない軟水を使うと本来の日本茶の香味が味わえます。サーバーに氷を入れ、水を注ぐだけで氷水が出来ますが、炻器（せっき）の湯冷ましを使うと日本らしい雰囲気を同時に演出出来、結露で濡れた陶器も見栄えが美しいでしょう。

First of all, prepare ice made from soft water without any smell of chlorine to bring out the most of the tea's natural flavor. Put the ice in a small pitcher and pour cold water on top of it. Any vessel can be used as long as it can handle cold temperatures and has a spout. However, by using stoneware vessels you will be able to create a Japanese atmosphere, and the condensation also adds a subtle beauty to experience.

2

茶葉を10～15g急須に入れます。これぐらい茶葉をたっぷり使うと、熱いお湯で淹れると苦味と渋味が強く出て美味しくありませんが、氷水ならうま味と甘味が引き出されて贅沢な味わいになります。

Put 10 to 15g of *Sencha* in the teapot. Using this much leaves when steeping tea in hot water will make it extremely bitter and astringent, but iced water will bring out only the *umami* and sweetness and turn it into an extremely rich taste experience.

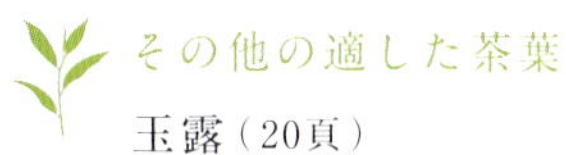

その他の適した茶葉

玉露（20頁）

Other suitable teas

Gyokuro (p.20)

淹れ方	Steeping guide
茶葉の量　10 ~ 15g	Tea leaves　10 ~ 15g
湯量　20ml	Water　20ml
湯温　氷水	Temperature　Iced water
時間　180秒	Steeping time 180 seconds

3

茶葉を茶さじで均一に広げます。茶葉を壊さないようにやさしくならします。一般的な形状の急須だと茶葉の山が出来、浸らない茶葉も出てしまいますので、均一に広げられる平型の急須が適しています。

Spread out the tea leaves evenly with the tea measure. Make sure to do it gently so as not to crush any of the tea leaves. A normal teapot with a small bottom will result in a heap of tea leaves that will not be immersed, so use a teapot with a flat bottom. These are called *Hiragata Kyusu* in Japanese.

4

氷水を注ぎます。ちょうどお茶の葉が浸るほどの量が目安です。たくさん注ぎすぎず、全ての茶葉を浸すために円を描くように少しずつ注ぎます。

Pour iced water into the pot, but only until the leaves are barely soaked. To avoid pouring too much water, pour a little by little as if you were drawing a circle.

｜煎茶｜氷茶の淹れ方

Steeping Sencha in iced water

5

氷水出し茶は凝縮されたうま味を味わうのがポイントですが、湯呑みを温めておくと香りも楽しむことが出来ます。火傷に注意しながら湯呑みに熱湯を注ぎます。

Steeping tea in iced water is a lot about enjoying *umami*, but preheating the small cups will make the aroma enjoyable as well. Make sure not to burn yourself when you pour hot water into the cups.

6

湯呑みに入ったお湯を湯こぼしなどに捨てます。残った水分がお茶を薄め、お湯と濃いお茶が混ざると違和感を感じることもありますので、水滴を丁寧に拭きましょう。

Discard the hot water in the cups. The tea is supposed to have a very rich taste but any remaining water drops will weaken the tea and give it an odd taste, so make sure to wipe the cups thoroughly.

7

３分経ってから湯呑みの縁に触れるように静かに注ぎます。温めてある湯呑みに触れると香りが立ち上ぼりやすくなります。出来上がったお茶はほんの数滴ですが、香味が爆発的に口中で広がりますので、ゆっくりと味わいましょう。

After three minutes, pour slowly and quietly along the edge of the cup. When the tea comes in contact with the hot surface of the cup, the aroma rises easily. You will only end up with a few drops of tea, but the taste and flavor will unfold explosively in your mouth, so sip it slowly, little by little.

二煎目、三煎目を味わう

一煎目でほとんどのうま味が出たので、二煎目はキレのある渋味と香りを引き出すために熱めで淹れます。ただ、茶葉の量が多いため苦味・渋味が出やすいので、待たずにすぐ湯呑みに注ぐのがポイントです。三煎目は湯冷ましを使わず直接ケトルから注いでも良いですが、すぐに注ぐと雑味が出ずにあっさりしたお茶が味わえます。

How to enjoy the second and third steeping

Most of the *umami* is extracted in the first steeping, so use hot water to extract a pleasant astringency and more of the aroma. However, since a lot of leaves are used in this case, bitterness and astringency dissolve easily so make sure not to wait but pour the tea immediately in the cups. For the third steeping, you can pour straight from the kettle, but again make sure to keep the steeping time down to only a few seconds to fully enjoy a light and refreshing third steeping without any excess bitterness.

｜煎茶｜ボトル冷茶の淹れ方
How to make cold brewed Sencha

「蒸し製緑茶」は冷たいお水でも美味しく淹れられます
Japanese tea makes an excellent beverage even when steeped in cold water

1

まず、茶漉しが付いているボトルを用意します。茶葉を15 ～ 20gほど入れてから冷水を750cc注ぎます。日本の水なら基本的には何でも構いませんが、海外では硬水の地域が多く、その場合は日本茶の本来の香味を引き出すために軟水器で濾過した水かボトリングされた軟水を使いましょう。

First of all, prepare a bottle that has a strainer. Put 15 - 20g of tea leaves in the bottle and pour about 750 ml of cold water. If you happen to be in Japan, most water is suitable for Japanese tea, but if the tap water in your area is hard, use a water softener or purchase bottled soft water.

2

冷水はお湯と比べて浸出がゆっくり進み、表面にも茶葉が浮いてきます。浸出が均一に進むように、蓋をした後、何度かボトルを上下にかえして茶葉を水に馴染ませてから冷蔵庫に入れます。夜に支度し、寝ている間に浸出させるのが味と香りを手軽に出せる方法ですが、2時間ほどでもそれなりに香味を楽しめます。

It takes time to extract the taste and flavor in cold water, and some leaves tend to float on the surface. To bring out the best of the tea, tip the bottle upside down a couple of times and mix the leaves thoroughly before refrigerating. If you prepare it before going to bed, both the taste and flavor will be extracted to the maximum while you sleep. Otherwise at least 2 hours will result in a fairly good cold brewed *Sencha*.

その他の適した茶葉

深蒸し煎茶（19頁）/玉露（20頁）/
かぶせ茶（21頁）/ほうじ茶（24頁）
茎茶・棒茶（25頁）

Other suitable teas

Fukamushi Sencha (p.19)
Gyokuro (p.20) / Hojicha (p.24)
Kukicha, Bocha (p.25)

淹れ方	Steeping guide
茶葉の量　15 - 20g	Tea leaves　15 - 20g
水量　750cc	Water　750ml
浸出時間　最短2時間 （最適な時間は8時間程度）	Steeping time　2 h (About 8 hours if possible)

3

冷蔵庫から取り出したら、茶葉が沈み、上下で液体の濃さに違いが生じているため、支度した時と同じように何度か上下にかえします。この時に茶葉と水色の変化の様子が明確にわかるのがガラスボトルの利点です。

When taking it out of the fridge, the tea leaves will have sunk to the bottom making the clear layer on top paler and weaker in taste compared to the slightly cloudy liquor in the bottom. To even it out, turn it upside down a few times just like when you were preparing it. To be able to see how the leaves and the color of the liquor change at this point is clearly one of the advantages of using a glass bottle.

4

冷茶ボトルはあくまでも大きめのガラスの急須のようなもので、茶殻が残ると渋味や雑味などが出る原因になります。急須を使う時と同じように、他のボトルまたはデキャンターなどに注ぎ切ります。すぐに全部をサーブしない場合は冷蔵庫に保管して24時間以内に消費しましょう。

A bottle for making cold brewed tea is after all nothing but a big glass teapot, so when the tea leaves remain for too long, harshness and excess astringency will inevitably enter the scene. To avoid this, pour everything into another bottle or decanter just like you would pour all your tea into teacups from a teapot. If you do not intend to drink everything at once, keep the remaining tea in the fridge and make sure to consume it within 24 hours.

｜煎茶｜ボトル冷茶の楽しみ方
How to enjoy cold brewed Sencha

ワイングラスで…

意外かもしれませんが、冷茶の香りを味わうにはワイングラスが最も適しています。ワイングラスは低めの温度の香りが楽しめるように作られたものですから、固定観念に縛られずに是非とも試してみてください。冷茶は色やキャラクターが白ワインに似ているので、グラスは白ワイン用が最適です。この飲み方なら香りだけでなく、エレガントな雰囲気も合わせて演出できます。「お昼のワイン」として味わってみると、きっと新しいマリアージュも生まれるでしょう。

Pouring it into a wine glass...

It might sound surprising, but the best way to enjoy the aroma of cold brewed *Sencha* is to drink it from a wine glass. A wine glass is designed to enjoy the aroma even at low temperatures, so getting rid of prejudices and use it for cold tea as well is something that everyone should try. White wine glasses, which will actually make the tea resemble wine, are the most suitable type for the purpose. In this way you can enjoy not only the aroma but also create a special atmosphere of elegance. Trying it out as a "wine for the lunch" may well prove to be an unexpected way to come up with great new food pairing.

玉露の淹れ方 | How to steep Gyokuro

玉露の本来の香味を最も引き出した味わい方
How to bring out the most of the characteristic taste of Gyokuro

1 玉露は濃いうま味を最大限に味わえるように低温で淹れます。まずは、お湯を冷ますために湯冷ましを2個準備します。

In order to fully enjoy the *umami* of *Gyokuro*, it should be steeped at low temperatures. First of all, prepare two *yuzamashi* (water coolers) to cool down the water.

2 茶葉は宝瓶(写真参照)または小ぶりの急須に入れて均一に広げます。お湯はケトルから1つ目の湯冷ましに注ぎます。

Put the leaves in a small teapot or a small flat "*hohin*" (pictured) and spread them evenly. Pour hot water from the kettle into the first *yuzamashi*.

3 湯冷ましを持つとまだ熱く感じるので、さらに冷ますために2つ目の湯冷ましに注ぎます。新しい器に移すと、湯温は約10℃下がります。

The water in the *yuzamashi* will not only feel but actually be very hot, so in order to cool it down even more, pour the water into the second *yuzamashi*. Whenever you pour from one vessel to another the temperature drops about 10°C.

4 しばらく待ち、湯冷ましを手で持っても熱く感じなくなれば約50℃です。温度の調整が出来たら静かに注ぎます。

After waiting for a while, the *yuzamashi* will not feel overly hot even if you hold it. When the temperature has reached around 50°C (120°F), pour it gently into the teapot.

淹れ方	Steeping guide
茶葉の量　6g	Tea leaves　6g
湯量　60cc	Water　60ml
湯温　50度	Temperature　50°C (120°F)
時間　120秒	Steeping time　120 seconds

5　蓋をして2分ほどじっくりと浸出させます。注ぐ時には、濃さを均一にするためにそれぞれの湯呑みに少しずつ廻し注ぎします。
うま味が凝縮された玉露は少量で味わうため、小ぶりの「豆茶碗」などを使います。二煎目も美味しく味わえるように最後の一滴まで注ぎ切ります。

Put on the lid and steep the tea for about two minutes. Pour it little by little into each cup and keep repeating this to adjust the strength.
Since *Gyokuro* is very thick and rich in *umami*, it is best enjoyed in small quantities from tiny cups. Make sure to pour all the liquor into the cups in order not to ruin the second steeping.

二煎目を味わう

二煎目以降は徐々に湯温を上げるため、一度だけ湯冷ましをして注ぎます。一煎目はうま味を楽しみ、二煎目からは玉露の独特な香りと甘い余韻が味わえます。

How to enjoy the second steeping

For the second and following steepings, raise the temperature of the water gradually. In the first steeping *umami* plays the main part, whereas in the second and beyond it will weaken in favor of the unique aroma and sweet aftertaste.

茶葉を食べる | Eating tea leaves

美味しい日本茶は食べても美味しいです。特に玉露は渋味が少なく、茶殻が柔らかいため三煎目を淹れてからポン酢や醤油をかけて食べると茶葉に閉じ込められている香味も含めて味わえます。

Good Japanese tea will also taste good when eaten. Especially *Gyokuro*, with its soft leaves low in astringency, can be turned into a true delicacy after the third steeping. By pouring *Ponzu* (a type of Japanese citrus juice) or soy sauce on top of the used leaves, you will be able to enjoy the taste and flavor trapped in the leaves in a completely different way.

抹茶の点て方 | How to whisk Matcha

茶筅で点てた抹茶は、本来の香味が引き出せる最も美味しい味わい方
Using a chasen to whisk matcha is the best way to bring out the natural flavor and taste

1

まずは点てる前の準備として抹茶をふるいで濾します。ついつい省いてしまいがちな作業ですが、点てる時に塊が残らないようにするための大事なステップです。

Before whisking, start by sifting the *matcha*. This step is often overlooked by many, but very important in order to avoid lumps when whisking.

2

一度、茶碗にお湯を注ぎ、茶筅も濡らします。こうすることで茶筅を清められ、同時に穂先が柔らかくなり、抹茶が点てやすくなります。

Pour hot water into the bowl, and soak the *chasen* (bamboo whisk). By doing this, you are able to both rinse the *chasen* and make the tines softer, which makes it easier to whisk the *Matcha*.

点て方	Preparation guide	
抹茶の量　2g	Matcha powder	2g
湯量　100cc	Water	100ml
湯温　80度	Temperature	80°C (175°F)

3

抹茶は決まった点て方をしなければならないと思われがちですが、煎茶と同様、基本的な点て方をベースに、好みに合わせて茶量、茶種、湯量、湯温、そして器や点て方で味や香りの変化を楽しめるものでもあります。例えば、薄茶に関しては、泡を多く点てることで、口当たりをやわらかくし、渋味も淡く抑えることが出来ます。泡を多く点てたい場合は、手首を利かせて泡立て器で泡を点てるように茶筅を振ります。良質な抹茶本来の味と香りを存分に楽しみたければ、泡を控えることで渋味とうま味をしっかり引き出せます。

There is a tendency to think that there is a correct way of whisking *Matcha*. However, just like *Sencha*, the taste and flavor changes depending on the amount of tea, type of *Matcha*, temperature, and of course the tea bowl and how the tea is whisked. All these factors can be adjusted to suit one's own preference. For example, in the case of *usucha* (weak *Matcha*), whisking the tea so as to produce a fine froth will weaken the astringency and give the tea a smooth mouthfeel. If this is your goal, you should use your wrist to move the *chasen* back and forth in a similar manner to when you are whisking other liquids. However, if you want to enjoy high quality *Matcha* to the maximum, you can extract both astringency and *umami* fully by mixing the *Matcha* carefully, without making it frothy.

ほうじ茶の淹れ方 | How to steep Hojicha

芳ばしい香りで食後などの時にすっきりします
Sweet roasted aroma - perfect after meals

1

まずは急須に茶葉を入れます。ほうじ茶らしい香ばしさとあっさり感をを引き出すために煎茶などよりも茶葉を少なめに計ります。煎茶と比べて軽いので、4gの場合、茶さじ2杯、またはティースプーンで4杯ほどを目安にします

First, put the tea leaves in the teapot. To bring out the most of the characteristic aroma and lightness, use fewer leaves than when you are steeping *Sencha*. Since the leaves are lighter than *Sencha*, take two full Japanese tea measures or roughly four teaspoons to measure four grams.

2

急須に入れた葉を均一に広げます。少ないように見えるかもしれませんが、入れすぎると雑味が出る原因になります。

Spread the tea leaves evenly in the teapot. It might look like a small amount, but too much leaves will only bring out excess bitterness.

その他の適した茶葉

番茶（22頁）/京番茶（23頁）
粉茶（26頁） ※ティーバッグを使用

Other suitable teas

Bancha (p.22) / Kyo-bancha (p.23)
Konacha (p.26) *Using the tea bag

淹れ方		Steeping guide	
茶葉の量	4g	Tea leaves	4g
湯量	200cc	Water	200ml
湯温	熱湯	Temperature	Boiling water
時間	30秒	Time	30 seconds

3

お湯を注ぎます。日本茶は一般的にお湯を冷ましてから淹れますが、ほうじ茶の場合、特徴となる香りは熱湯を使った方が楽しめます。

Pour boiling water into the teapot. In general, the water is cooled down after boiling when steeping Japanese tea. However, in the case of *Hojicha* the characteristic aroma is more easily extracted and better enjoyed when steeped at a high temperature.

4

長く待ち過ぎると雑味が出るため、30秒ほどで濃さを調整しながら湯呑みに注ぎ分けます。短い浸出時間に違和感を感じるかもしれませんが、ほうじ茶の香りを引き出すにはこれで十分です。

Oversteeping will give the tea a harsh and bitter taste so start pouring after 30 seconds. Pour little by little to make the strength even. 30 seconds might sound short, but is enough to extract the characteristic aroma of *Hojicha*.

Tea in plastic bottles
ペットボトルのお茶について

初めて日本に来た外国人は日本人の緑茶の飲み方に驚くでしょう。誰もが丁寧に急須でお茶を淹れているかと思いきや、ペットボトルから直接飲むことが最も一般的な飲み方だからです。烏龍茶と紅茶に次いで、ペットボトルの緑茶飲料の製造に初めて成功したのは1985年のこと。その時に商品化された伊藤園の「お～いお茶」はまさに日本人のお茶の飲み方を変えた存在となりました。開発にあたっては難点がいくつかありましたが、緑茶が紅茶や烏龍茶と違って非常に変質しやすいことが最も大きなハードルでした。

変質しやすい一番茶ではなく、硬化が進んだ二番茶以降の茶葉を使い、さらに安定させるためにアスコルビン酸（ビタミンC）を加えることで解決しました。これで砂糖など甘味料が入っていない日常的な緑茶飲料が生まれました。使用されている茶葉から考えると現代版の番茶として捉えられ、どこでもお茶が飲めるという意味で多くの日本人にとってはありがたい存在です。

便利なペットボトル茶の普及によって急須で淹れたお茶の魅力が忘れられているのではないかと心配する人もいるかもしれませんが、急須で淹れる消費者が少なくなってきた背景にはいくつかの要因があります。かつての日本では祖父母の世代と一緒に暮らすことが一般的でしたが、核家族や共働きが進んだため、家族と一緒にお茶を飲む場面が減ってしまいました。加えてコーヒーや紅茶、炭酸飲料などの普及により、嗜好品の市場が多様化しました。このような要因を鑑みると、ペットボトルのお茶が登場しなければ、現代社会においてお茶と触れる機会はもっと少なくなっていたかもしれません。

急須で淹れた緑茶が好きな人には違和感があるかもしれませんが、ペットボトルという形で緑茶が大量に消費されていることは、近代化してもやはり日本人は緑茶を捨てたくないと思っている証拠の一つでしょう。

Foreigners visiting Japan for the first time are probably surprised when they see how the Japanese drink green tea. Many probably picture all Japanese carefully steeping their tea in traditional teapots, only to find out that most drink their tea from plastic bottles. After canned and bottled versions of oolong tea and black tea, the green tea equivalent was finally introduced in 1985. The product, "Oi Ocha" by Itoen, truly came to change the way the Japanese consumed green tea. During the development stage there were a lot of hurdles, but the most challenging one was that the quality of steeped green tea alters more easily than black or oolong tea.

In particular *Ichibancha* (first harvest tea) is prone to change after steeping, compared to the more stable later harvests which were chosen as the raw material. In addition, ascorbic acid (Vitamin C) was added as a stabilizer and this proved to be the solution. In this is way, a ready to drink tea beverage without any added sugar or other sweeteners was successfully created.

Considering the type of tea used for making the product, tea in plastic bottles could be interpreted as a modern version of *Bancha*, greatly valued by many Japanese since it can be enjoyed anywhere. Some are worried that the convenience of bottled tea has made people forget the joy of steeping tea with teapots. In fact, there are many factors behind the lesser use of teapots among modern day Japanese. In the past, several generations used to live under the same roof, but with the household structure changing in favor of nuclear families and with more working women in society, tea drinking with the family has, to a large degree, disappeared. On top of that, the introduction of coffee, black tea and various carbonated beverages has diversified the market, thereby making green tea vulnerable to competition. Considering all these factors, it is possible that natural opportunities to come in contact with green tea would be even fewer in modern society if bottled tea did not exist.

Although devoted tea drinkers who use teapots might consider tea in plastic bottles a blasphemy, its very existence can also be thought of as evidence that the Japanese are not willing to give up on tea despite society getting modernized.

急須の楽しみ

The beauty of Japanese Teapots

急須のセレクション | A selection of Japanese teapots

急須の選び方 | How to choose a teapot ?

好みの色やデザインなどで急須を選ぶことももちろん大事ですが、お茶を淹れるための道具なので、使い勝手とお茶を美味しく淹れられることが大前提です。炻器ならお茶の苦味・渋味をある程度吸着するので味はまろやかになります。吸水性のある陶器はお茶の香りを奪ってしまうので高品質な茶に使うのは適切ではありません。磁器は、苦味・渋味の吸着が極めて少ないのですが、熱の伝わり方が早く、湯温をコントロールするのに工夫が必要です。私の経験では常滑焼や万古焼の精度が高い炻器の急須であれば基本的に美味しく淹れられるように思います。

Choosing a teapot with a good design is of course important, but since it is a tool made for steeping tea, functionality and good tasting tea is the major premise. Stone ware teapots absorb bitterness and astringency to some degree, which makes the tea taste milder. Very porous ceramic teapots, on the other hand, will absorb even the aroma and are therefore not suitable for Japanese tea. White porcelain is not necessarily bad, but since it does not absorb bitterness or astringency to any notable degree, it tends to give the tea a slightly sharp taste. From my experience, steeping Japanese tea in well crafted stoneware pots from either Tokoname or Yokkaichi (Bankoyaki) in principle always gives a good result.

金属網 | Metal strainer

陶器網 | Ceramic strainer

素材の次に見るのが茶漉しです。細かい陶器製の茶漉しの方が注ぎやすいのですが、深蒸し煎茶や粉茶などのような細かい茶葉の場合は茶漉しが詰まってしまうため、金属製の広い網や籠が入った急須が良いでしょう。他にも茶種によって適した急須が異なります。例えば、少量で低温で淹れる玉露なら小さめの急須か宝瓶が使いやすく、逆にほうじ茶のように熱くてたっぷり淹れるお茶は保温性が高い厚手の器の方が熱さと香りを引き出せます。

After the material, the next thing to check is the tea strainer. A ceramic strainer with many small holes makes it easier to pour tea, but is unsuitable when steeping teas with small particles such as *Fukamushi Sencha* and *Konacha* since they tend to clog the strainer. For those teas it is better to use a teapot with a metal strainer. In the case of *Gyokuro*, which is steeped in small quantities at low temperatures, a small teapot or a so called *hohin* is suitable. Tea steeped at high temperatures like *Hojicha*, on the other hand, requires a thick teapot that will preserve the heat so that there is no unnecessary loss of aroma.

お茶は見るものにあらず、飲むものです — 黒の茶器セット

上質な日本茶は薄手の白い磁器の湯呑みで色を見ながら味わうのが基本ですが、それは中国から伝わってきた習慣です。お茶は見るものにあらず、飲むものです。その概念を最大限に演出した画期的な炻器の茶器セットが登場しました。従来の茶器と全く違い、お茶の色がはっきり見えないようにあえて黒にされています。そしてずっしりと重くつくり、手に取ると器だけでなく、お茶自体にも存在感を感じます。抹茶の場合、茶道という日本特有の喫茶文化がありますが、煎茶道を見ると大陸の影響が目立ちます。近代に入ってようやく煎茶にも日本らしさを感じる喫茶文化が形成されてきました。それをさらに進化させるかのように、新しい味わいを生み出したこの常滑の茶器セットは21世紀の茶を楽しむゲームチェンジャーになるかもしれません。

"Tea is for drinking, not for looking at"— Tokoname black tea set

To drink tea from thin white porcelain is by many considered to be the standard. This way of enjoying tea has its origin in China, just like tea itself. However, tea is something that we enjoy more with our palate rather than with our eyes. This thought is expressed to its utmost in this new kind of black stoneware tea set that saw daylight only recently. Contrary to conventional thinking, the color black is chosen on purpose so that the color of the liquor cannot be perceived, and this makes it completely different from other tea ware. Furthermore, the heaviness of the cups resembles that of Raku tea bowls, and holding them highlights the presence, not only of the utensils, but also the tea itself. In the world of *Matcha* we find the tea ceremony, a peculiar Japanese cultural practice. However in the case of *Sencha* ceremonies, continental influence clearly stands out. In modern days, we have finally seen a genuinely Japanese tea drinking culture take form. By bringing to life a completely new way of enjoying Japanese tea, this Tokoname tea set seems push this even further. It might very well be a game changer.

ブレケル・オスカルのセレクト茶について
SENCHAISM, Oscar Brekell's Tea Selection

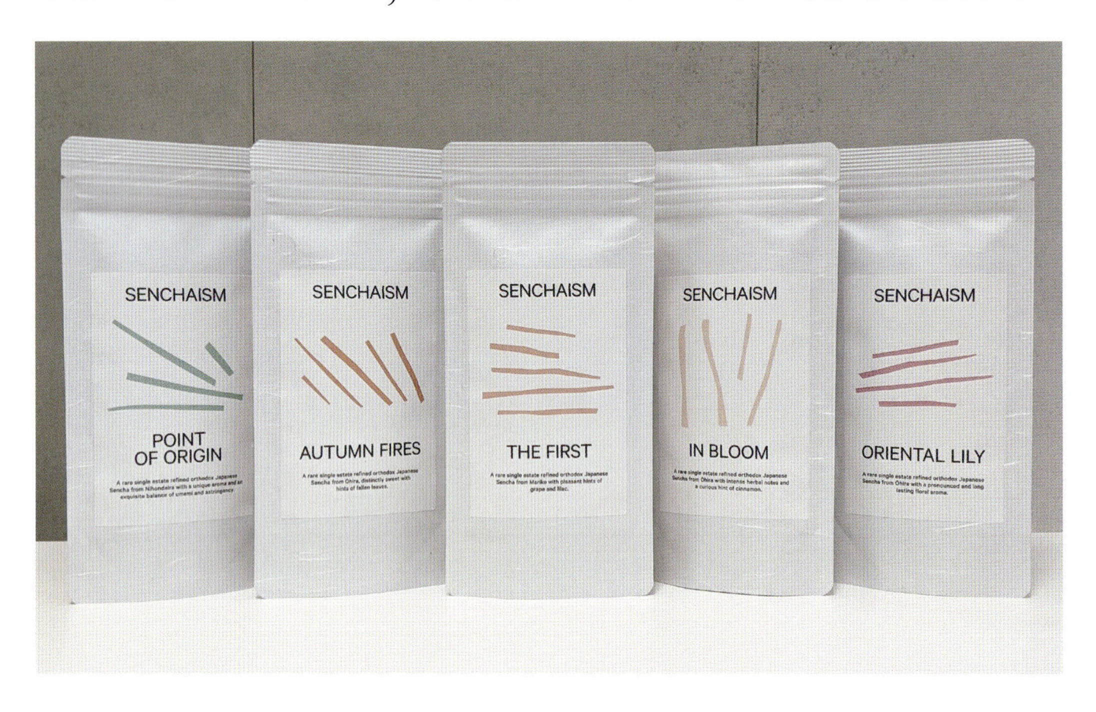

日本には美味しいお茶がたくさんありますが、私が「最も奥が深くて面白い」と考えているのはシングルオリジンの日本茶です。この本でも紹介しているように（38～45頁）同じ煎茶でも品種によって花の香り、ハーブの香り、フルーティな香り、そしてストレートに山の香りがする品種があります。うま味や渋味などもそれぞれ変わりますから、まさにワインやシングルモルトのように個性豊かな味わいです。これらの単一品種・単一農園だけで仕上げられたお茶は20年ほど前に登場してまだ新しいため、生産量は少ないのですが、日本茶の可能性を強く感じる特別なお茶です。

お茶の時間を大事にしている私は、このような日本茶を最も味わい深いと感じます。そこで、より多くの方の生活が豊かになるよう、シングルオリジンを中心とした日本茶ブランド「SENCHAISM」を立ち上げました。日本の方だけでなく、国籍を問わず多くの方に味わっていただきたいお茶です。

Japan is abundant with good tea, but what I find most intriguing and fascinating is single estate Japanese tea. As introduced in this book (see p.38-45), *Sencha* can taste very different, having everything from floral, herbal, fruity or forest-like notes all depending on the cultivar. The degree of taste elements like *umami* and astringency also differs and, being this rich in variation, *Sencha* can really be enjoyed like a fine wine or a single malt whiskey. Refined single estate and single cultivar Japanese tea came into existence only about twenty years ago, and thus the production volumes are still comparably small. However, this relatively new phenomenon clearly shows the potential of Japanese tea, and this makes it truly special. For a person like myself, who greatly values my teatime, without doubt I feel like this type of tea is by far the most enjoyable. Therefore I started the brand "SENCHAISM", putting emphasis on single estate Japanese Tea, with a wish to enrich more people's lives. My hope is that this will be enjoyed not only by the Japanese, but by tea lovers from all corners of the world.

Glossary

A

Aracha (unrefined tea) 32, 34

B

Bancha 22

C

Catechin 13, 15

Chasaji (tea scoop) 66

Chasen (bamboo whisk) 29, 82

Chakan (tea canister) 66

Chazutsu (tea canister) 66

F

Fukamushi Sencha ("Deep steamed Sencha") 19

Fukuoka (tea growing region) 62

Futaoki (lid rest) 66

G

Gifu (tea growing region) 61

Gogumi (blending) 36

Goishicha 31

Gyokuro 20

H

Hiragata Kyushu (flat teapot) 72

Hojicha 24

Honyama (tea growing region) 53

I

Ichibancha 64

K

Kabusecha 21

Kagoshima (tea growing region) 63

Kakegawa (tea growing region) 55

Kamairicha (Japanese pan-fired tea) 27

Kawane (tea growing region) 55

Konacha 26

Kukicha 25

Kyobancha 23

Kyoto (tea growing region) 56

Kyusu (teapot) 66

L

L-Theanine (umami compound) 13

M

Makinohara (tea growing region) 54

Matcha 29, 82

Mie (tea growing region) 59

N

Nagatani Soen (the father of Sencha) 51

Nara (tea growing region) 59

Nibancha 64

Nishio (tea growing region) 61

S

Sanbancha 64

Sayama (tea growing region) 60

Sencha (orthodox) 18

Shiagecha (refined tea) 34

Shiga (tea growing region) 58

Shikoku (tea growing region) 62

Shincha 14

Shizuoka (tea growing region) 52

Single estate Japanese tea 38

T

Tamaryokucha 27

Temomicha (hand-rolled Japanese tea) 35

Tencha 28

U

Uji (tea growing region) 57

Umami 13

W

Wakocha (Japanese black tea) 30

Y

Yunomi (teacup) 66

Yuzamashi (Hot water cooler) 66

Z

Zairai (Seed-propagated plant) 40

あとがき

高校の頃に日本茶にはまった私は、情報を探し求めた過程で100年以上前に岡倉天心によって書かれた「The book of tea」に巡り会いました。岡倉天心は、西洋人は東洋の文化を積極的に理解しようとしなくても、東洋のものであるお茶だけを完全に受け入れたと述べました。グローバル化が進んできた今の世界は当時とだいぶ違って異文化交流が出来る機会が増えましたが、お茶とは表面的な付き合いにしか至っていない西洋人がまだ多いでしょう。距離と言葉という壁が立ちはだかっていますが、日本茶への関心が世界的に高まりつつある中で新しい日本茶の本に挑むべき時代がきたとここ数年痛感しました。

日本茶は単なる嗜好品ではなく、長い歴史と奥が深い関連文化もあって非常に多面的な存在です。そして国籍を問わずに人と人を繋ぐ不思議な力も持っています。異文化に対しての理解が深まり、摩擦を和らげる意味で平和な世界を作ることにも繋がると私は信じています。魅力に溢れた日本茶のことを２ヶ国の言葉で表現することにより、ようやく表面的な付き合いを卒業してより多くの人が本当の意味で日本茶と豊かな付き合いが出来ることを願っています。

最後になりますが、「バイリンガル日本茶BOOK」を手にとって頂き、まことにありがとうございます。この本が、多面的で奥が深い日本茶の楽しい世界への扉になれたなら幸いに存じます。参考にして頂きながら、日本茶との自分らしい触れ方にぜひ辿り着いてほしいと思います。

私と同じように、あなたの生活も日本茶で豊かになることを心より祈っております。

Afterword

During my high school years, I became interested in Japanese tea, and in my search for information I came across "The book of tea" by Kakuzo Okakura, a book written over a hundred years ago in which he explains the Eastern mindset by using tea culture as a pivot. Okakura states that, although Westerners show little interest in Eastern culture, they have wholeheartedly accepted the Eastern practice of drinking tea. Today's world, where globalization is progressing rapidly and intercultural exchange has become more common, is very much different from the world in which Okakura lived. Even so, perhaps many Westerners are still enjoying tea in a rather superficial way. However, despite the distance and the language barrier, which make up great hurdles, more and more people across the globe are becoming interested in Japanese tea. Because of this, I have for many years strongly felt that the time is ripe for a new book in the field.

Japanese tea is not just a simple beverage. It is extremely multifaceted with its long history and all the complex cultural practices that surround it. Furthermore, tea brings people together, regardless of nationality. In the sense that it deepens the understanding of other cultures and thereby lessens the friction between us, I believe that tea drinking may help us build a more peaceful world. By attempting to put the beauty of Japanese tea into words in two languages, I sincerely hope that more people will graduate from superficial tea drinking and fully enjoy more of the wonder it has to offer.

Finally, I would like to thank you for taking your time reading "The Book of Japanese Tea". If this book could work as a gateway to the wonderful world of one of the most fascinating and complex beverages there is, I could not be happier. I also hope that this will inspire people and lead them to enjoy Japanese tea in a way that suits themselves. From the bottom of my heart, I hope that Japanese tea will enrich your life, just as it has enriched mine.

プロフィール　ブレケル・オスカル

1985年、スウェーデン生まれ。学生時代に日本茶に魅せられ来日。2010年、岐阜大学に留学。その後、日本企業に就職して2013年に再来日。2014年、日本茶インストラクターの資格を取得。静岡県茶業研究センターでの研修生や日本茶輸出促進協議会での職員などを経て、2018年に独立し、現在は海外での日本茶教育などに携りつつ国内外でお茶の魅力を伝える講座やセミナーなどを開催し、日本茶の普及につとめる。外国人初の手揉み茶の教師補の資格も持つ。2016年、世界緑茶協会の「CHAllenge」賞を受賞。

Profile　Per Oscar Brekell

Born in Sweden in 1985, Per Oscar Brekell developed an interest in Japanese tea during high school, something that grew into a passion that later led him to relocate to Japan. In 2010, he studied at Gifu University and in 2013 he came back to Japan for a job at a Japanese company. He became a certified Japanese Tea Instructor in 2014, completed an internship at The Tea Research Center in Shizuoka, and has also been working for Japan Tea Export Council. In 2018 he set up his own business, and is now involved in tea education projects overseas and arranges tea events and seminars in Japan. He is the first non-Japanese to receive a certificate for making hand-made *Sencha* and in 2016 he was awarded the CHAllenge prize by the World Green Tea Association.

デザイン　久都間ひろみ［くつま舎］
写真　大道雪代
石部健太朗
本杉勇人
上田葵
ブレケル・オスカル
池田奈実子［農研機構果樹茶業研究部門］
静岡県茶業研究センター
掛川市
鹿児島県茶業会議所
井ノ倉光博［井ノ倉ティーファーム］
吉野亜湖
山形蓮
山政小山園
編集協力　石部健太朗
吉野亜湖
校正　石部健太朗
吉野亜湖
神長健二
Marzi Pecen
Justus Wallen
Noli Ergas

Design　Hiromi Kutsuma［Kutsumasha］
Photos　Yukiyo Daido
Kentaro Ishibe
Hayato Motosugi
Aoi Ueda
Per Oscar Brekell
Namiko Ikeda
［National Institute of Fruit Tree and Tea Science, Japan］
Shizuoka Prefecture Tea Research Center
Kakegawa city
Tea Industry Chamber of Kagoshima
Mitsuhiro Inokura［TEA FARM INOKURA］
Ako Yoshino
Ren Yamagata
Yamamasa-Koyamaen
Editorial Assistance　Kentaro Ishibe
Ako Yoshino
Proofing　Kentaro Ishibe
Ako Yoshino
Kenji Kaminaga
Marzi Pecen
Justus Wallen
Noli Ergas

ブレケル・オスカルの
バイリンガル日本茶BOOK

2018年9月13日　初版発行
2022年4月27日　2版発行

著　者　ブレケル・オスカル
発行者　納屋嘉人
発行所　株式会社　淡交社
本　社　〒603-8588 京都市北区堀川通鞍馬口上ル
［営業］Tel. 075-432-5156［編集］Tel. 075-432-5161
支　社　〒162-0061 東京都新宿区市谷柳町39-1
［営業］Tel. 03-5269-7941［編集］Tel. 03-5269-1691
www.tankosha.co.jp
印刷・製本　亜細亜印刷株式会社

ISBN978-4-473-04261-3

The Book of Japanese Tea
By Per Oscar Brekell

This book was published in 2018
by Tankosha Publishing Co., Ltd.